DATE DUE

DE 10 04			

DEMCO 38-296

UMTS Networks
Architecture, Mobility and Services

UMTS Networks
Architecture, Mobility and Services

Heikki Kaaranen
Aqua Records Oy, Finland

Ari Ahtiainen
Nokia Research Center, Finland

Lauri Laitinen
Nokia Research Center, Finland

Siamäk Naghian
Nokia Networks Oy, Finland

Valtteri Niemi
Nokia Research Center, Finland

JOHN WILEY & SONS, LTD

Chichester • New York • Weinheim • Brisbane • Singapore • Toronto

...nd

...7

... service enquiries): cs-books@wiley.co.uk
...vww.wiley.co.uk or http://www.wiley.com

Other Wiley Editorial Offices

John Wiley & Sons, Inc., 605 Third Avenue,
New York, NY 10158-0012, USA

WILEY-VCH Verlag GmbH
Pappelallee 3, D-69469 Weinheim, Germany

John Wiley & Sons Australia Ltd, 33 Park Road, Milton,
Queensland 4064, Australia

John Wiley & Sons (Canada) Ltd, 22 Worcester Road
Rexdale, Ontario, M9W 1L1, Canada

John Wiley & Sons (Asia) Pte Ltd, 2 Clementi Loop #02-01,
Jin Xing Distripark, Singapore 129809

British Library Cataloguing in Publication Data

A catalogue record for this book is available from the British Library

Image on front cover was supplied by Xplorer (www.xplorer.co.uk).

ISBN 0471 48654 X

Typeset in Times by Deerpark Publishing Services Ltd, Shannon.
Printed and bound in Great Britain by Biddles Ltd, Guildford and King's Lynn.
This book is printed on acid-free paper responsibly manufactured from sustainable forestry, in which at least two trees
are planted for each one used for paper production.

Contents

Preface

The world's first public GSM call was made on 1st July 1991 in a city park of Helsinki, Finland. That event is now regarded as the birthday of the second generation mobile telephony. GSM has been an overwhelming success, which was difficult to predict at that early stage. In the past 10 years GSM has become a truly global system for mobile communications. We now have cellular phone penetration rates exceeding 50% in many countries and approaching 80% in the Nordic countries.

Ten years later GSM has brought us onto the footstep of the third generation mobile communications system – Universal Mobile Telecommunications System (UMTS). Again the first networks are being opened and a new generation of fancy mobile phones will appear. By the end of February 2001 already 61 UMTS licenses were awarded to network operators in 16 different countries.

UMTS networks will introduce into wide area use a completely new high bit rate radio technology – wideband CDMA (Code Division Multiple Access). On the other hand the core network part of the UMTS system is firmly founded on the successful GSM network, which has evolved from circuit switched voice network into a global platform for mobile packet data services like short messaging, mobile web browsing and mobile e-mail access.

The latest estimates show, that packet switching traffic in the mobile core networks will exceed circuit switching by 2005. This transition is enabled by the UMTS system, which makes it possible for the network operators to provide equally strong circuit-switched and packet-switched domains to meet the speed and capacity demands. Most voice and time-critical data services may still use circuit switching, while less time-sensitive data passes through the UMTS mobile packet core network.

One of the key promises made by the UMTS mobile computing and communications devices is the ability to deliver information to users almost anytime and anywhere. In UMTS the mobile phone is becoming a personal trusted device, a life management tool for work and leisure. Among the new possibilities for communication, entertainment and business are new kinds of rich call and multimedia data services, which are fuelled by the mobility and personalisation of users and their terminals.

This is a book about UMTS networks as a third generation platform for mobility and services. It aims to provide a comprehensive overview of the system architecture and its evolution and to serve as a guidebook to those who need to study the specifications from the Third Generation Partnership Program (3GPP). The content of the book is divided into three parts.

The first part consists of Chapters 1 and 2, which serve as an executive summary of the

UMTS system. Chapter 1 introduces the UMTS technical and service architecture and key system concepts. Chapter 2 is an illustrated story about mobile network evolution from second generation GSM into the first UMTS release and beyond that towards full-IP mobility networks.

The second part consists of Chapters 3–8, which examine the radio access and core network in more detail, explaining the functions and services provided to the end users. Chapter 3 on UMTS radio communications provides the fundamentals of cellular radio which are necessary for understanding the WCDMA radio access system as part of the UMTS network. Chapters 4 and 5 present the functional split and system management aspects distributed among the UMTS network elements in the radio access and core network parts. Chapter 6 provides an overview on the UMTS user equipment focusing on those aspects, which are most visible to the rest of the UMTS network. In Chapter 7 the UMTS network is examined as a network for services. The UMTS service capabilities (WAP, CAMEL, location based services, etc.) are first introduced as a value-added service platform and then application examples with different Quality of Service (QoS) characteristics are discussed. The advanced security solutions of the UMTS network are then discussed in Chapter 8.

The remaining Chapters 9 and 10 form the third part of the book. In these chapters we take a protocol-oriented view describing the system-wide interworking between the different architectural elements. Chapter 9 first elaborates the basic UMTS protocol architecture and then introduces the individual system protocols one by one. Chapter 10 returns to the network-wide view by showing selected examples of system procedures, which describe how the transactions are carried out across the UMTS network interfaces under the co-ordination of the system protocols.

At the footstep of the third generation mobile communications the success of the UMTS will come from the thousands of leading system and software engineers, content providers, application developers, system integrators and network operators. We hope this book will help all of them to reach their targets and let them enjoy and benefit from the UMTS networking environment.

This book represents the views and opinions of the authors and does not necessarily represent the views of their employers.

Acknowledgements

During the "UMTS Book project" the authoring team members have had a pleasure to follow the exciting finalisation of the UMTS system specifications. Among the colleagues both from Nokia and outside, those listed below have provided valuable input and comments on the various aspects of the book: Seppo Alanara, Mika Forssell, Harri Holma, Kaisu Iisakkila, Tatjana Issayeva, Pekka Korja, Jan Kåll, Juho Laatu, John Loughney, Atte Länsisalmi, Anna Markkanen, Tomi Mikkonen, Juha Mikola, Ahti Muhonen, Mikko Puuskari, Mikko J. Rinne, Ville Ruutu, Juha Sipilä, Janne Tervonen, Mikko Tirronen, Ari Tourunen and Jukka Vialén.

This inspiring working environment and close contacts with the R&D and standardisation programs within Nokia were made possible by the following managers in those programs: Kari Aaltonen, Heikki Ahava, Tapio Harila, Jari Lehmusvuori, Juhani Kuusi, Yrjö Neuvo, Pertti Paski, Tuula-Mari Rautala, Tuomo Sipilä, Jukka Soikkeli, Jari Vainikka and Asko Vilavaara.

The publishing team lead by Mark Hammond at John Wiley & Sons Ltd. has participated in the production of this book by giving us excellent support. Their hard working spirit made it possible to keep the demanding schedule in the publication process.

After all, this is a book about UMTS networks, which are based on the joint design and engineering effort of numerous colleagues of ours, who have brought their expertise together to make it happen. Without being able to list all the experts from the early 1990s until the recent meetings of the 3GPP organisation, we like to thank all of them for their dedicated work.

Finally we like to express loving thanks to all the members of our families for the patience and support shown during the long days and late nights of the book writing effort. Among them, Mrs Satu Kangasjärvelä-Kaaranen deserves special thanks, since her help in word processing and graphical design of numerous figures was invaluable in putting together the manuscript.

The authors welcome any comments and suggestions for improvements or changes that could be implemented in possible new editions of this book. The e-mail address for gathering such input is umtsnetworks@pcuf.fi.

Helsinki, Finland.
The authors of *UMTS Networks*.

Abbreviations

1G	First Generation
2G	Second Generation
3G	Third Generation
3GPP	Third Generation Partnership Project
4G	Fourth Generation
8-PSK	Octagonal Phase Shift Keying
A	
AAA	Authorisation, Authentication and Accounting
AAL	ATM Adaptation Layer
ABR	Available Bit Rate service
AC	Admission Control, also Authentication Centre
ACK	Acknowledge
ACM	Address Complete Message (ISUP)
ACTS	Advanced Communications Technologies and Services
AH	Authentication Header
AICH	Acquisition Indicator Channel
AK	Anonymity Key
AKA	Authentication and Key Agreement
AM	Acknowledged Mode in RLC
AMF	Authentication Management Field
AMPS	American Mobile Phone System
AMR	Adaptive Multi Rate
ANS	Answer Message (ISUP)
AOA	Angle of Arrival Positioning
AP	Application Part
API	Application Programming Interface
APN	Access Point Name
ARIB	Association of Radio Industries and Business
AT	Attention Command
ATD	Absolute Time Difference
ATM	Asynchronous Transfer Mode
AuC	Authentication Centre
AUTN	Authentication token
AUTS	Authentication Synchronisation

B

BCCH	Broadcast Control Channel
BCFE	Broadcast Control Function Entity
BCH	Broadcast Channel
BER	Bit Error Rate
BIB	Backward Indication Bits
B-ISDN	Broadband ISDN
BMC	Broadcast/Multicast Control
BRAN	Broadband Radio Access Network
BS	Base Station
BSN	Backward Sequence Number
BSS	Base Station System
BTS	Base Transceiver Station (GSM)

C

CA	Certificate Authorities
CA-ICH	CPCH Channel Assignment Indicator Channel
CAMEL	Customised Application for Mobile Network Enhanced Logic
CAP	CAMEL Application Part
CBR	Constant Bit Rate Service
CC	Call Control
CC	Country Code
CCCH	Common Control Channel
CCPCH	Common Control Physical Channel
CD-ICH	CPCH Collision Detection Indicator Channel
CDMA	Code Division Multiple Access
CDMA2000	A CDMA System in North America
CDR	Call Detail Record
CF	Call Forwarding
CFN	Connection Frame Number
CGI	Cell Global Identity
CI	Cell Identity
CIC	Circuit Identification Code
CK	Cipher Key
CLP	Cell Loss Priority
CLPC	Closed Loop Power Control
CM	Communication Management
CN	Core Network
COMC	Communication Control
CP	Control Plane
CPC	Centralised Power Control
CPCH	Common Packet Channel
CPICH	Common Pilot Channel
CRC	Cyclic Redundancy Check
CRNC	Controlling Radio Network Controller
C-RNTI	Cell Radio Network Temporary Identity
CS	Circuit Switched

CS	Convergence Sublayer (ATM)
CSCF	Call Server Control Function
CSICH	CPCH Status Indication Channel
CSPDN	Circuit Switched Public Data Network
CTCH	Common Traffic Channel
Cu	Interface between TE and USIM
CW	Call Waiting
CWTS	China Wireless Telecommunication Standard Group
D	
DB	Decibel
DCCH	Dedicated Control Channel
DCFE	Dedicated Control Function Entity
DCH	Dedicated Channel
DES	Data Encryption Standard
DGPS	Differential Global Positioning System
DL	Downlink
DoS	Denial of Service
DPCCH	Dedicated Physical Control Channel
DPDCH	Dedicated Physical Data Channel
DRNC	Drift Radio Network Controller
DRNS	Drift RNS
D-RNTI	Drift RNC Radio Network Temporary Identifier
DRX	Discontinuous Reception
DS-CDMA	Direct Sequence CDMA
DSCH	Downlink Shared Channel
DSPC	Destination Signalling Point Code
DTCH	Dedicated Traffic Channel
DTE	Data Terminal Equipment
E	
E2E	End-to-End
EC	Echo Cancelling Equipment
ECT	Explicit Call Transfer
EDGE	Enhanced Data Rates for GSM Evolution
E-GPRS	Enhanced GPRS
EFR	Enhanced Full Rate
E-HSCSD	Enhanced HSCSD
EIR	Equipment Identity Centre
EIRP	Equivalent Isotropic Radiated Power
E-RAN	EDGE Radio Access Network
ESP	Encapsulation Security Payload
ETR	ETSI Technical Report
ETS	ETSI Telecommunication Standard
ETSI	European Telecommunications Standards Institute
F	
FACH	Forward Access Channel
FBI	Feedback Information

FCCH	Frequency Correction Channel
FDD	Frequency Division Duplex
FDMA	Frequency Division Multiple Access
FER	Frame Error Rate
FH-CDMA	Frequency hopping CDMA
FIB	Forward Indication Bits
FISU	Fill-in Signalling Unit
FP	Frame Protocol
FPLMTS	Future Public land Mobile Telephony System
FRAMES	ACTS Future Radio wideband Multiple Access System
FSN	Forward Sequence Number
G	
Gb	GPRS Interface Between SGSN and GSM BSS
Gc	Interface Between GGSN and HLR/AuC
GERAN	GSM/EDGE Radio Access Network
Gf	Interface Between SGSN and EIR
GGSN	Gateway GPRS Support Node
Gi	Interface Between GGSN and External Network
GMLC	Gateway Mobile Location Centre
GMM	GPRS Mobility Management
GMSC	Gateway MSC
Gn	Interface Between Two GSNs
Gp	Interface Between Two GGSNs
GPRS	General Packet Radio Service
GPS	Global Positioning System
Gr	Interface Between SGSN and HLR/AuC
Gs	Interface Between SGSN and Serving MSC/VLR
GSM	Global System for Mobile Communications
GSMS	GPRS SMS
gsm SCF	GSM Service Control Function
GSN	GPRS Support Nodes
GTD	Geometric Time Difference
GTP	GPRS Tunnelling Protocol
GTP-C	GTP for Control Signalling
GTP-U	GTP for User Plane
Gx	Any G Interface
H	
H.323	A ITU-T Protocol
HCS	Hierarchical Cell Structure
HDR	Header
HEC	Header Error Control
HFN	Hyper Frame Number
HLR	Home Location Register
HM-CDMA	Hybrid Modulation CDMA
HO	Handover
HON	Handover Number

HSCSD	High Speed Circuit Switched Data
HSPA	High Speed Packet Access
HSS	Home Subscriber Server
HTML	HyperText Markup Language
HTTP	HyperText Transfer Protocol
HW	Hardware
Hz	Hertz, Cycles per Second
I	
I/Q	In-phase/Quadrature
IAM	Initial Address Message (ISUP)
ICC	Integrated Circuit Card
ICGW	Incoming Call Gateway
ID	Identifier
IDEA	International Data Encryption Algorithm
IEC	International Electrotechnical Commission
IETF	Internet Engineering Task Force
IK	Integrity Key
IKE	Internet Key Exchange
IMEI	International Mobile Equipment Identity
IMS	IP Multimedia Subsystem
IMSI	International Mobile Subscriber Identity
IMT-2000	International Mobile Telecommunications 2000
IN	Intelligent Network
IP	Internet Protocol
IPDL	Idle Period Down Link
IPSEC	IP Security protocol
IPSP	IP Signalling Point
IPv4	Internet Protocol Version 4
IPv6	Internet Protocol Version 6
IS-95	North American Version of the CDMA Standard
ISCP	Interference Signal Code Power
ISDN	Integrated Services Digital Network
ISO	International Organisation for Standardisation
ISUP	ISDN User Part
ITU	International Telecommunication Union
ITU-T	ITU Telecommunication Standardisation Sector
Iu	UMTS Interface Between 3G-MSC/SGSN and RNC
Iub	UMTS Interface Between RNC and BS
Iur	UMTS Interface Between RNCs
K	
KAC	Key Administration Centre
kb/s	Kilo-bits per Second
kHz	Kilohertz
L	
L1	Layer 1 – Radio Physical Layer
L2	Layer 2 – Radio Data Link Layer

L3	Layer 3 – Radio Network Layer
LA	Location Area
LAC	Location Area Code
LAI	Location Area Identity
LAN	Local Area Network
Lc	Interface between GMLC and GSMSCF
LCS	Location Services
LCS	Location Communication System
Lg	Interface Between GMLC and MSC/SGSNs
Lh	Interface Between GMLC and HSS
LI	Length Indicator
LMU	Position Measurement Unit
LOS	Line of Sight
LSSU	Line Status Signalling Unit
M	
M3UA	MTP3 User Adaptation Layer
MAC	Medium Access Control
MAC	Message Authentication Code
MAP	Mobile Application Part
MAPSEC	MAP Security Protocol
Mb/s	Megabits per Second
MCC	Mobile Country Code
MC-CDMA	Multi-carrier CDMA
Mcps	Megachips per Second
MCU	Multipoint Control Unit
MD5	Message Digest #5
ME	Mobile Equipment
MEHO	Mobile Evaluated Handover
MExE	Mobile Execution Environment
MGCF	Media Gateway Control Function
MGW	Media Gateway
MHz	Megahertz
MIB	Marter Information Block
MM	Mobility Management
MNC	Mobile Network Code
MO	Mobile Originated
MOBC	Mobility Control
MOC	Mobile Originated Call
MoNet	RACE Mobile Networks Project
MP3	MPEG 1 Audio Layer 3
MPEG	Moving Pictures Expert Group
MRC	Maximum Ratio Combining
MRF	Media Resource Function
MS	Mobile Station
MSC	Mobile Switching Centre
MSE	MExE Service Environment

MSISDN	Mobile Subscriber ISDN Number
MSN	Mobile Subscriber Number
MSRN	Mobile Station Roaming Number
MSU	Message Signalling Unit
MT	Mobile Terminated
MT	Mobile Termination
MTC	Mobile Terminated Call
MTP	Message Transfer Part
MTP3	Message Transfer Part Level 3
MTU	Maximum Transmission Unit
N	
NAS	Network Access Server
NAS	Non-Access Stratum
NBAP	Node B Application Part
NDC	National Destination Code
NEHO	Network Evaluated Handover
NLOS	Non-Line Of Sight
NMS	Network Management Subsystem
NMT	Nordic Mobile Telephone
NRT	Non-Real Time
NSS	Network Sub System
NT	Network Termination
Nt	Notification (SAP)
NW	Network
O	
O&M	Operations and Maintenance
OHG	Operator Harmonisation Group
OLPC	Open Loop Power Control
OQPSK	Offset Quadrature Phase Shift Keying
OSA	Open Service Architecture
OSI	Open System Interconnection
OSPC	Originating Signalling Point Code
OTDOA	Observed TDOA
P	
PC	Power Control
PCCH	Paging Control Channel
P-CCPCH	Primary Common Control Physical Channel
PCH	Paging Channel
PCM	Pulse Code Modulation
PCPCH	Physical Common Packet Channel
PCS	Personal Communication System
PD	Protocol Discriminator
PDC	Pacific Digital Communication
PDCP	Packet Data Convergence Protocol
PDH	Plesiochronous Digital Hierarchy
PDP	Packet Data Protocol (e.g. PPP, IP, X.25)

PDSCH   Physical Downlink Shared Channel
PDU   Packet Data Unit
PDU   Protocol Data Unit
PEM   Privacy Enhanced Mail
PGP   Pretty Good Privacy
PHS   Personal Handyphone System
PHY   Physical layer
PICH   Page Indicator Channel
PIN   Personal Identification Number
PKI   Public Key Infrastructure
PLMN   Public Land Mobile Network
PN   Pseudo Noise
PNFE   Paging and Notification Function Entity
POC   PSTN Originated Call
PPP   Point-to-Point Protocol
PRACH   Physical Random Access Channel
PS   Packet Switched
PSC   Primary Synchronisation Code
P-SCH   Physical Shared Channel
PSPDN   Packet Switched Public Data Network
PSTN   Public Switched Telephone Network
PT   Payload Type
PTC   PSTN Terminated Call
P-TMSI   Packet TMSI
Q
QoS   Quality of Service
QPSK   Quadrature Phase Shift Keying
R
R   Interface Between TE and MT
R4   Release 4 of 3GPP UMTS Standard
R5   Release 5 of 3GPP UMTS Standard
R99   Release 1999 of 3GPP UMTS Standard
RA   Routing Area
RAB   Radio Access Bearer
RACE   Research in Advanced Communications in Europe
RACH   Random Access Channel
RADIUS   Remote Authentication Dial-In User Service
RAI   Routing Area Identity
RAN   Radio Access Network
RANAP   Radio Access Network Application Part
RAND   Random Number (Used for Authentication)
RAS   Registration, Admission, and Status
RAU   Routing Area Update
RB   Radio Bearer
REL   Release Message (ISUP)
RES   Response in Authentication

RF	Radio Frequency
RFC	Request for Comments in IETF
RFE	Routing Functional Entity
RL	Radio Link
RLC	Radio Link Control
RLC	Release Complete Message (ISUP)
RLCP	Radio Link Control Protocol
RNBP	Reference Node Based Positioning
RNC	Radio Network Controller
RNS	Radio Network Subsystem
RNSAP	Radio Network Subsystem Application Part
RNTI	Radio Network Temporary Identity
ROHC	Robust Header Compression
RRC	Radio Resource Control
RRM	Radio Resource Management
RSA	Public-key Security Algorithm by Rivest, Shamir and Adleman
RT	Radio Termination
RT	Real Time
RTCP	Real-time Control Protocol
RTD	Relative Time Difference
RTP	Real-time Transport Protocol
RTT	Round-Trip Time Based Positioning
S	
SA	Security Association
SAI	Service Area Identifier
SAP	Service Access Point
SAR	Segmentation and Re-assembly Sublayer
SB	Scheduling Block
SCCH	Synchronisation Control Channel
SCCP	Signalling Connection Control Part
S-CCPCH	Secondary Common Control Physical Channel
SCE	Service Creation Environment
SCF	Service Control Function
SCH	Synchronisation Channel
SCI	Subscriber Controlled Input
SCP	Service Control Point
SCTP	Stream Control Transport Protocol
SDH	Synchronous Digital Hierarchy
SDU	Service Data Unit
SET	Secure Electronic Transactions
SFN	System Frame Number
SGSN	Serving GPRS Support Node
SHA-1	Secure Hash Algorithm #1
S-HTTP	Secure Hypertext Transfer Protocol
SIB	System Information Block
SIF	Service Information Field

SIM	GSM Subscriber Identity Module
SIO	Service Information Octet
SIP	Session Initiation Protocol
SIR	Signal-to-Interference Ratio
SLIP	Serial Line Internet Protocol
SLS	Signalling Link Selection
SM	Session Management
S-MIME	Secured Multipurpose Internet mail Extension
SMLC	Serving Mobile Location Centre LC
SMS	Short Message Service
SMSC	Short Message Service Centre
SN	Sequence Number
SN	Serving Network
SN	Subscriber Number
SOCKS	Socket Security
SPC	Signalling Point Code
SQN	Sequence Number
SRB	Signalling Radio Bearer
SRNC	Serving Radio Network Controller
SRNS	Serving RNS
S-RNTI	SRNC Radio Network Temporary Identity
SS	Supplementary Service
SS7	Signalling System No. 7
SSCF	Service Specific Co-ordination Function
SSCF-NNI	Service Specific Co-ordination Function – Network Node Interface
S-SCH	Secondary Synchronisation Channel
SSCOP	Service Specific Connection Oriented Protocol
SSCOP	Service-Specific Connection Oriented Protocol
SSDT	Site Selection Diversity Transmission
SSL	Secure Socket Layer
STM	Synchronous Transfer Module
STP	Signalling Transfer Point
SW	Software
SYSINFO	System Information
T	
TA	Terminal Adaptation
TBS	Transport Block Set
TCAP	Transaction Capabilities Application Part
TCP	Transmission Control Protocol
TDD	Time Division Duplex
TDMA	Time Division Multiple Access
TDOA	Time Difference of Arrival positioning
TE	Terminal Equipment
TEID	Tunnel Endpoint Identifier
TF	Transport Format
TFC	Transport Format Combination

TFCI	Transport Format Combination Indicator
TFI	Transport Format Indicator
TH-CDMA	Time Hopping CDMA
TI	Transaction Identifier
TLS	Transport Layer Security
TMSI	Temporary Mobile Subscriber Identity
TOA	Time-of-Arrival Positioning
TPC	Transmit Power Control
TR	Technical Report (3GPP,ETSI)
Tr	Transparent Mode in RLC
TRAU	Transcoding and Rate Adaptation Unit
TRX	Transceiver
TS	Technical Specification (3GPP,ETSI)
TTA	Telecommunication Technology Association
TTC	Telecommunication Technology Committee
TTP	Traffic Termination Point
Tu	Interface between NT and RT
U	
UBR	Unspecified Bit Rate Service
UDP	User Datagram Protocol
UE	User Equipment
UICC	Universal Integrated Circuit Card
UL	Uplink
Um	Radio Interface for GSM BSS
UM	Unacknowledged Mode in RLC
UMTS	Universal Mobile Telecommunications System
UP	User Plane
URA	UTRAN Registration Area
URL	Uniform Resource Locator
U-RNTI	UTRAN Radio Network Temporary Identity
U-RNTI	UTRAN Radio Network Temporary Identifier
USAT	UMTS SIM Application Toolkit
USIM	Universal Subscriber Identity Module
UTRA	Universal Terrestrial Radio Access
UTRAN	Universal Terrestrial Radio Access Network
Uu	radio interface for UTRA
V	
VAS	Value Added Service Platform
VBR	Variable Bit Rate Service
VCI	Virtual Circuit Identifier
VHE	Virtual Home Environment
VLR	Visitor Location Register
VMS	Voice Mail System
VoIP	Voice Over IP
VPI	Virtual Path Identifier
VSC	Videotext Service Centre

W

W	Watt
WAP	Wireless Application Protocol
WARC	World Administrative Radio Conference
WCDMA	Wideband Code Division Multiple Access
WLAN	Wireless Local Area Network
WML	Wireless Markup Language
WTLS	Wireless Transport Layer Security
WWW	World Wide Web

X

X.25	An ITU-T Protocol for Packet Switched Networks
X.509	Internet X.509 Public Key Infrastructure
XMAC	Expected Message Authentication Code
XRES	Expected user Response

1

Introduction

Nowadays it is widely recognised that there are three different generations as far as mobile communication is concerned (Figure 1.1). The first generation, 1G, is the name for the analogue or semi-analogue (analogue radio path, but digital switching) mobile networks established in the mid 1980s such as the Nordic Mobile Telephone system (NMT) and American Mobile Phone System (AMPS). These networks offered basic services for the users and the emphasis was on speech and speech-related services. 1G networks were developed with national scope only and very often the main technical requirements were agreed between the governmental telecom operator and domestic industry without wider publication of the specifications. Due to the national specifications the 1G networks were incompatible with each other and mobile communication was considered to be some kind of curiosity and added value service on top of the fixed networks in those times.

Because the need for mobile communication increased, also the need for a more global mobile communication system increased. The international specification bodies started to

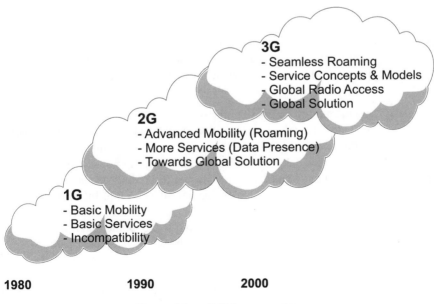

Figure 1.1. Cellular generations

specify what the second generation, 2G, mobile communication system should look like. The emphasis on 2G was on compatibility and international transparency; the system should be regional (like European-wide) or a semi-global one and the users of the system should be able to access it basically anywhere within the region. From the end-users point of view, 2G networks offered a more attractive "package" to buy; besides the traditional speech service these networks were able to provide some data services and more sophisticated supplementary services. Due to the regional nature of standardisation, the concept of globalisation did not succeed completely and there are some 2G systems available on the market. Out of these, the commercial success story is Global System for Mobile Communications (GSM) and its adaptations: it has clearly exceeded all the expectations set both technically and commercially.

The third generation, 3G, is expected to complete the globalisation process of mobile communication. Again there are national and regional interests involved and difficulties can be foreseen. Anyway the trend is that 3G will mostly be based on GSM technical solutions due to two reasons: the GSM technology dominates the market and great investments made in GSM should be utilised as much as possible. Based on this, the specification bodies created a vision about how mobile telecommunication will develop within the next decade. Through this vision, some requirements for 3G were short-listed as follows:

- The system must be fully specified (like GSM) and major interfaces should be standardised and open. The specifications generated should be valid world-wide.
- The system must bring clear added value to the GSM in all aspects. However, in the beginning the system must be backward compatible at least with GSM and ISDN (Integrated Services Digital Network).
- Multimedia and all of its components must be supported throughout the system.
- The radio access of the 3G must provide wideband capacity be generic enough in order to become available world-wide. The term "wideband" was adopted to reflect the capacity requirements between 2G narrowband capacity and the broadband capacity of the fixed communications media.
- The services for end-users must be independent from radio access technology details and the network infrastructure must not limit the services to be generated. That is, the technology platform is one issue and the services using the platform are totally another issue.

While the 3G specification work is still going on the major telecommunication trends have changed, too. The traditional telecommunication world and up to now the separate data communications or the Internet have started to converge rapidly. This has started a development chain, where traditional telecommunication and IP technologies are combined in the same package. This common trend has many names depending on the speaker's point of view; some people call the target of this development the 'Mobile Information Society' or "Mobile IP", some other people say it is "3G All IP" and in some commercial contexts also the name "E2E IP" (End-to-End IP) is used. From a 3G point of view, a full-scale IP implementation is defined as one targeted phase of the 3G development path.

The 3G system is therefore already in evolution through new phases and recently the discussion on 4G has already started. Right now it may be too early to predict where the 3G evolution ends and 4G really starts. Rather this future development can be thought of as an ongoing development chain where 3G will continue to introduce new ways to handle and combine all kinds of data and mobility. 4G will then emerge as a more sophisticated system concept bringing still more added value to the end-users.

1.1   Specification Process for 3G

The uniform GSM standard in European countries has enabled globalisation of mobile communications. This became evident, when the Japanese 2G Pacific Digital Communications (PDC) failed to spread to the Far East and the open GSM standard was adopted on major parts of the Asian markets and when its variant became one of the nationally standardised alternatives for the US Personal Communication System (PCS) market, too.

A common, global mobile communication system naturally creates a lot of political desires. In the case of 3G this can be seen even in the naming policy of the system. The most neutral term is third generation, 3G. In different parts of the world different issues are emphasised and thus the global term 3G has regional synonyms. In Europe 3G has become UMTS (Universal Mobile Telecommunication System), following the ETSI perspective. In Japan and the US the 3G system often carries the name IMT-2000 (International Mobile Telephony 2000). This name comes from the International Telecommunication Union (ITU) development project. In the US the CDMA2000 is also one aspect of 3G cellular systems and it represents the evolution from the IS-95 system. In this book, we will describe the UMTS system as it has been specified by the world-wide 3G Partnership Project (3GPP).

In the beginning UMTS inherited plenty of elements and functional principles from GSM and the most considerable new development is related to the radio access part of the network. UMTS brings into the system an advanced access technology, namely wideband type of radio access. Wideband radio access is implemented with Wideband Code Division Multiple Access (WCDMA) technology. WCDMA is evolved from CDMA, which, as a proven technology, has been used for military purposes and for narrowband cellular networks especially in the US.

The UMTS standardisation was preceded by several pre-standardisation research projects founded and financed by the EU. During the years 1992–1995 a RACE MoNet project developed the modelling technique describing function allocation between the radio access and core parts of the network. This kind of modelling technique was needed, for example, in order to compare Intelligent Network (IN) and GSM MAP protocols as mobility management solutions. This was, besides the discussion on the broadband vs narrowband ISDN, one of the main dissents in MoNet. Also the discussions about use of ATM (Asynchronous Transfer Mode) and B-ISDN as fixed transmission techniques arose in the end of the MoNet project.

During the following years 1995–1998 3G research activities were continued within the ACTS FRAMES project. The first years were used for selecting and developing a suitable multiple access technology, considering mainly the TDMA (Time Division Multiple Access) versus CDMA (Code Division Multiple Access). The big European manufacturers preferred TDMA because it was used also in GSM. CDMA based technology was promoted mainly by US industry, which had experiences with this technology mainly due to its early utilisation in defence applications.

ITU had a dream to specify at least one common global radio interface technology. This kind of harmonisation work was done under the name Future Public Land Mobile Telephony

System (FPLMTS) and later IMT-2000. Due to many parallel activities in regional standardisation bodies this effort turned into promotion of common architectural principles among the family of IMT-2000 systems.

Europe and Japan also had different short-term targets for 3G system development. In Europe a need for commercial mobile data services with guaranteed quality, for example mobile video services, was widely recognised after the early experiences from narrowband GSM data applications. Meanwhile in the densely populated Far East there was an urgent demand for additional radio frequencies for speech services. The frequency bands identified by ITU in 1992 for the future 3G system called IMT-2000 became the most obvious solution to this issue. In early 1998 a major push forward was achieved when ETSI TC-SMG decided to select WCDMA as its UMTS radio technology. This was also supported by the largest Japanese operator NTT DoCoMo. The core network technology was at the same time agreed to be developed on the basis of the GSM core network technology. During 1998 the European ETSI and the Japanese standardisation bodies (TTC and ARIB) agreed to make a common UMTS standard. After this agreement, the 3GPP organization was established (Figure 1.2) and the determined UMTS standardisation was started worldwide.

From the UMTS point of view, the 3GPP organisation is a kind of "umbrella" aiming to

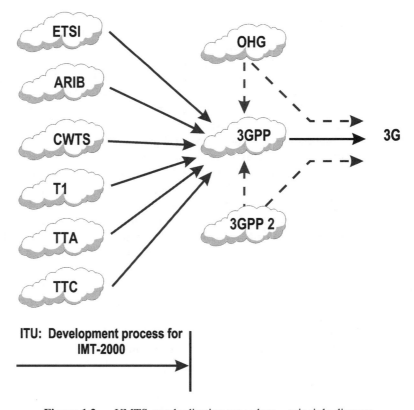

Figure 1.2.   UMTS standardisation procedure – principle diagram

form compromised standards by taking into account political, industrial and commercial pressures coming from the local specification bodies:

- ETSI (European Telecommunication Standard Institute)/Europe
- ARIB (Association of Radio Industries and Business)/Japan
- CWTS (China Wireless Telecommunication Standard group)/China
- T1 (Standardisation Committee T1 – Telecommunications)/US
- Telecommunication Technology Association (TTA)/Korea
- Telecommunications Technology Committee (TTC)/Japan

As this is a very difficult task an independent organisation called OHG (Operator Harmonisation Group) was established relatively soon after the 3GPP. The main task for 3GPP is to define and maintain UMTS specifications and the role of OHG is to look for compromise solutions for those items the 3GPP cannot handle internally. This arrangement guarantees that 3GPP's work will proceed on schedule.

To ensure that the American viewpoint will be taken into account a separate Third Generation Partnership Project Number 2 (3GPP-2) was founded and this organisation performs specification work from the IS-95 radio technology basis. The common goal for 3GPP, OHG and 3GPP-2 is to create specifications according to which a global cellular system having wideband radio access could be implemented.

To summarise, there was three different approaches towards the global cellular system, 3G. These approaches and their building blocks are, on a rough level, presented in Table 1.1.

Table 1.1 3G variants and their building blocks

Variant	Radio access	Switching	2G basis
3G (US)	WCDMA, EDGE, CDMA2000	IS-41	IS-95, GSM1900, TDMA
3G (Europe)	WCDMA, GSM, EDGE	Advanced GSM NSS and packet core	GSM900/1800
3G (Japan)	WCDMA	Advanced GSM NSS and packet core	PDC

When globality comes true, 3G specification makes it possible to take any of those switching systems mentioned in the table and combine them with any of the specified radio access parts and the result is a functioning 3G cellular network. The second row represents the European approach known as UMTS and this book gives an overview of its first release.

The 3GPP originally decided to prepare specifications on a yearly basis, the first specification release being Release 99 (3GPP R99). This first specification set has a relatively strong "GSM presence". From the UMTS point of view the GSM presence is very important; first, the UMTS network must be backward compatible with the existing GSM networks and second, the GSM and UMTS networks must be able to inter-operate together. The next release was known as 3GPP R00 but because of the multiplicity of changes proposed, the specification activities were scheduled into two specification releases 3GPP R4 and 3GPP R5. Also – to be consistent – the 3GPP R99 is sometimes called 3GPP R3. The 3GPP R4 defines major changes in UMTS core network circuit switched side and those are related to the separation of user data flows and their control mechanisms. Another big item in 3GPP R4 is to introduce mechanisms and arrangements for multimedia. An overview of 3GPP R4 is given in Chapter 5. The 3GPP R5 aims to introduce a UMTS network where the transport network utilises IP networking as much as possible. Therefore this goal is called the "All IP" network. IP and overlying protocols will be used in network control, too and also the user data flows are expected to be mainly IP based. In other words, the mobile network implemented according to the 3GPP R5 specifications will be an end-to-end packet switched cellular network using IP as the transport protocol instead of SS7 (System Signalling 7), which holds the major position in existing circuit switched networks. Naturally the IP-based network should still support circuit switched services, too. The 3GPP R4/R5 will also start to utilise the possibility for new radio access techniques. In 3GPP R99 the basis for the UMTS Terrestrial Access Network (UTRAN) is WCDMA radio access. In 3GPP R4/5 another radio access technology derived from GSM with Enhanced Data for GSM Evolution (EDGE) will be specified to create GSM/EDGE Radio Access Network (GERAN) as an alternative to building a UMTS mobile network.

1.2   Introduction to 3G Network Architecture

The main idea behind 3G is to prepare a universal infrastructure able to carry existing and also future services. The infrastructure should be designed so that technology changes and evolution can be adapted to the network without causing uncertainties to the existing services using the existing network structure. Separation of access technology, transport technology, service technology (connection control) and user applications from each other can handle this very demanding requirement. The structure of a 3G network can be modelled in many ways and here we introduce some ways to outline the basic structure of the network. The architectural approaches to be discussed in this section are:

- Conceptual network model
- Structural network architecture
- Resource management architecture
- UMTS service and bearer architecture

1.2.1   Conceptual Network Model

From the above-mentioned network conceptual model point of view, the entire network architecture can be divided into subsystems based on the nature of traffic, protocols structures, as well as physical elements. As far as the nature of traffic is concerned, the 3G network consists of two main domains, packet-switched (PS) and circuit-switched (CS) domains. According to the 3GPP specification TR 21.905 a *domain* refers to the highest-level group of physical entities and the defined interfaces (reference points) between such domains. The interfaces and their definitions describe exactly how the domains communicate with each other.

From the protocol structure and their responsibility point of view, the 3G network can be divided into two strata: access stratum and non-access stratum. *Stratum* refers to the way of grouping protocols related to one aspect of the services provided by one or several domains, see 3GPP specification TR 21.905. Thus, the access stratum contains the protocols handling activities between the User Equipment (UE) and access network. The non-access stratum contains protocols handling activities between the UE and Core Network (CS/PS domain), respectively. For further information about strata and protocols refer to Chapter 9.

The part of Figure 1.3 called "Home Network" maintains static subscription and security information. The serving network is the part of the core network + domain, which provides the core network functions locally to the user. The transit network is the core network part located on the communication path between the serving network and the remote party. If, for a given call, the remote party is located inside the same network as the originating UE, then no particular instance of the transit network is needed.

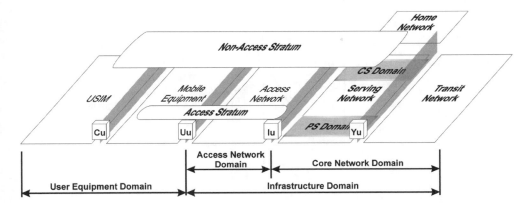

Figure 1.3.   UMTS architecture - conceptual model

1.2.2   Structural Network Architecture

In this book we mainly present the issues from the network structural architecture perspective. This perspective is presented in Figure 1.4. In UMTS the GSM technology has a remarkable role as background and actually UMTS aims to reuse everything, which is reasonable. For example, some procedures used within the non-access stratum are, in principle, reused from GSM but naturally with required modifications.

The 3G network terminal is called UE and it contains two separate parts, Mobile Equipment (ME) and UMTS Service Identity Module (USIM).

The new subsystem controlling the wideband radio access has different names, depending on the type of radio technology used. The general term is Radio Access Network (RAN). If especially talking about UMTS with WCDMA radio access, the name UTRAN or UTRA is used. The other type of RAN included in UMTS is GERAN. GERAN and its definitions are not part of 3GPP R99 though they are referred to as the possible radio access alternatives, which may be utilised in the future. The specification of GERAN and its harmonization with UTRAN is one of the topics in 3GPP R4 and 3GPP R5.

The UTRAN is divided into Radio Network Subsystems (RNS). One RNS consists of a set of radio elements and their corresponding controlling element. In UTRAN the radio element is Node B, referred to as Base Station (BS) in the rest of this book, and the controlling element is Radio Network Controller (RNC). The RNSs are connected to each other over access network-internal interface Iur. This structure and its advantages are explained in more detail in Chapter 4.

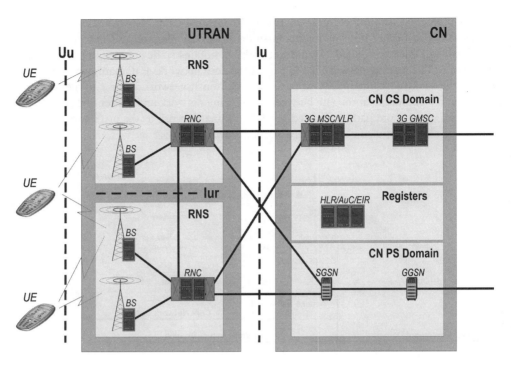

Figure 1.4. UMTS network architecture – network elements and their connections for user data transfer

The term Core Network (CN) covers all the network elements needed for switching and subscriber control. In early phases of UMTS part of these elements are directly inherited from GSM and they are modified for UMTS purposes. Later on, when transport technology changes, the core network internal structure will remarkably change, too. The term CN covers Circuit Switched (CS) and Packet Switched (PS) domains defined in Figure 1.3. The configuration alternatives and elements of the UMTS core network are discussed in Chapter 5.

The part of the Figure 1.4 called "Registers" is the same as Home Network in the preceding 3G network conceptual model. This part of the network maintains static subscription and security information. Registers are discussed in more detail in Chapter 5.

The major open interfaces of UMTS are also presented in Figure 1.4. Between the UE and UTRAN the open interface is Uu, which in UMTS is physically realised with WCDMA technology. For further details about WCDMA, refer to Chapter 3. The other major open interface is Iu located between the UTRAN and CN.

The RNSs are separated from each other with an open interface Iur. Iur is a remarkable difference when compared to GSM; it brings completely new abilities for the system to utilise so called macro diversity and also efficient radio resource management and mobility mechanisms. When the Iur interface is implemented in the network, the UE may attach to the network through several RNCs and each of those maintains a certain logical role during the radio connection. These roles are Serving RNC (SRNC), Drifting RNC (DRNC) and Controlling RNC (CRNC). The Controlling RNC has the overall control of the logical resources of its UTRAN access points, being mainly BSs. A SRNC is a role an RNC can take with respect to a specific connection between the UE and UTRAN. There is one SRNC for each UE that has a radio connection to UTRAN. The SRNC is in charge of the radio connection between the UE and the UTRAN. It also maintains the Iu interface to the core network, which is the main characteristic of the SRNC. A DRNC is a logical role used when radio resources of the connection between the UTRAN and the UE need to use cell(s) controlled by another but the SRNC itself. The UTRAN related issues in general are discussed in Chapter 4.

In addition to the CS and PS domains presented in Figure 1.4 the network may contain other domains. One example of these is the broadcast messaging domain, which is responsible for multicast messaging control. In this book we however concentrate on the UMTS network like that presented in Figure 1.4.

1.2.3 Resource Management Architecture

The network element–centric architecture described above results from functional decomposition and the split of responsibilities between major domains and finally between network elements. Figure 1.5 illustrates this split of major functionalities, which are

- Communication Management (CM)
- Mobility Management (MM)
- Radio Resource Management (RRM)

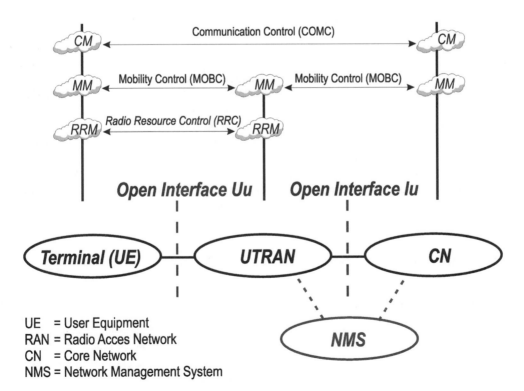

Figure 1.5. UMTS network architecture – management tasks and control duties

CM covers all of the functions and procedures related to the management of user connections. CM is divided into several sub-areas such as call handling for circuit switched connections, session management for packet switched connections, as well as handling of supplementary services and short message services. MM covers all of the functions and procedures needed for mobility and security, for example, connection security procedures and location update procedures. Most of the MM procedures occur within the core network and its elements but in the 3G part of the MM functions are performed also in UTRAN for packet switched connections. CM and MM principles are discussed in Chapter 5.

RRM is a collection of algorithms UTRAN uses for management of radio resources. These algorithms handle, for instance, power control for the radio connections, different types of handovers, system load and admission control. The RRM is an integral part of UTRAN and basic RRM is discussed more closely in Chapter 4. Some system-wide procedure examples about CM, MM and RRM functioning are given in Chapter 10.

Although these management tasks can be located within specific domains and network elements, they need to be supported by communication among the related domains and network elements. This communication is about gathering information and reporting about the status of remote entities as well as about giving commands to them in order to execute the management decisions. Therefore each of the management tasks is associated with a set of control duties such as:

- Communication Control (COMC)
- Mobility Control (MOBC)
- Radio Resource Control (RRC)

COMC maintains mechanisms like call control, and packet session control. MOBC maintains mechanisms, which cover, for example, execution control for location updates and security. The radio resources are completely handled within UTRAN and UE. The control duty called RRC takes care of, e.g. radio link establishment and maintenance between the UTRAN and UE. These collections of control duties are then further refined into a set of well-specified control protocols. For more detailed information about protocols, refer to Chapter 9.

As compared to the GSM system, this functional architecture has undergone some rethinking. The most visible change has to do with the mobility management, where responsibility has been split between UTRAN and CN. Also with regards to the RRM the UMTS architecture follows more strictly the principle of making UTRAN alone responsible for all radio resource management. This is underlined by the introduction of a generic and uniform control protocol for the Iu interface.

1.2.4 UMTS Service and Bearer Architecture

Both in 1G and 2G networks the technology implementation and its details were issues as such. In 3G the technology is important but the common opinion is that in 3G the network is more a "service network" than a plain cellular network. In other words, 1G and 2G were technologically limited networks allowing the end-users to use a limited set of technology-specific services. In 3G the technology should not be the limitation. This kind of approach is partially commercial and, in a way, gets the operator thinking. If this kind of service viewpoint is discussed, a 3G network can be modelled as shown in Figure 1.6.

Because a 3G network will be a very complex infrastructure as a whole, there are two

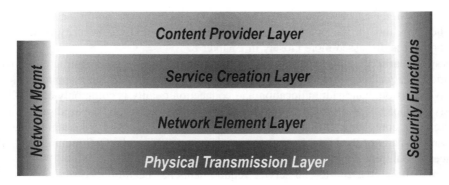

Figure 1.6.   UMTS network architecture – service model

issues, which are present everywhere in the network and should be handled carefully. These are network management and security. From the service architecture standpoint, a 3G network and its building elements can be divided into four different layers. The first layer being the basis of all is physical transmission. The nodes using physical transmission form their own layer, network elements. The third layer contains elements and functionalities forming the environment where the services for the end-users are created. Since the sophisticated services require contents, the content provision is separated into its own layer on top of the stack. It is often stated that the lower layers indicated in Figure 1.6 require the greatest investments and on the higher layers the investments are not playing the major role. Rather it can be said that every layer in this model generates both costs and revenue but those are different in nature. The two following principles can be applied to this four-layer stack:

- Rule A: the lower location the layer has, the bigger is the investment in network elements. In other words, transmission and network elements form the biggest cost in the 3G network. Revenue generation depends on the network configuration and coverage.
- Rule B: The higher location the layer has, the bigger is the investment in people and ideas. When moving to the higher layers, the pure technology is not an issue as such; more people having good ideas and an understanding of human behaviour are the core business principles here.

As a conclusion, a profitable and successful 3G network is a successful compromise and combination of those layers presented in Figure 1.6. In other words, technology forms a platform where good ideas have the possibility to turn into successful services for the end-users.

In traditional 2G networks the role of the network management is relatively limited and concerns normally the network elements and their technical behaviour. In 3G the idea of separating services and platforms and network widens the scope of network management; every service used requires involvement from every layer of this model and in order to ensure the correct functionality of the service the whole stack should be controlled.

The access level and network-internal security in UMTS is ensured by utilising very sophisticated mechanisms and effective algorithms. In this respect UMTS networks are more secure than GSM networks. The open business model supporting third party involvement brings in the aspect of end-to-end security. This is, however, not in the scope of 3GPP specifications though the potential risks involved are recognised. Security in the UMTS environment is discussed more closely in Chapter 8.

As stated earlier in this chapter, the 3G network mainly acts as an infrastructure providing facilities, adequate bandwidth and quality for the end-users and their applications. This facility provision, bandwidth allocation and connection quality is commonly called Quality of Service (QoS). If we think of an end-to-end service between users, the used service sets its requirements concerning QoS and this requirement must be met everywhere in the network. The various parts of the UMTS network contribute to fulfilling the QoS requirements of the services in different ways.

To model this, the end-to-end service requirements has been divided into three entities: local bearer service, UMTS bearer service and external bearer service. Local bearer service contains the mechanisms on how the end-user service is mapped between the terminal equipment and Mobile Termination (MT). Mobile termination is the part of the user equipment, which terminates the radio transmission to and from the network and adapts terminal equipment capabilities to those of the radio transmission. UMTS bearer service in turn contains mechanisms to allocate QoS over the UMTS/3G network consisting of UTRAN and CN. Since the UMTS network attaches itself to external network(s), the end-user QoS requirements must be handled towards the other networks, too. This is taken care of by the external bearer service.

Within the UMTS network the QoS handling is different in UTRAN and CN. From the CN point of view the UTRAN creates an "illusion" of a fixed bearer providing adequate QoS for the end-user service. This "illusion" is called radio access bearer service. Within the CN an own type of bearer service called CN bearer service is used. This division between radio

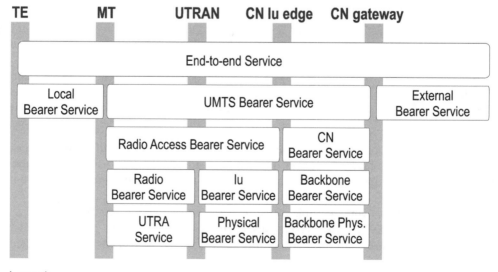

Figure 1.7.   Bearer architecture in UMTS

access bearer and CN bearer service is required since the QoS must be guaranteed in very different environments and both of these environments require their own mechanisms and protocols. For instance, CN bearer service is quite constant in nature since the backbone bearer service providing physical connections is also stable. Within UTRAN the radio access bearer experiences more changes as a function of time and UE's movement and this sets different challenges for QoS. This division pursues also the main architectural principle of the UMTS network, that is, the independence of the entire network infrastructure from the radio access technology.

The structure presented in Figure 1.7 is a network architecture model from the bearer and QoS point of view. Since the QoS is one of the most important issues in UMTS, the QoS and bearer concepts are handled throughout this book:

The rest of the book uses these architectural approaches as corner stones when exploring UMTS networks and their implementations.

2

Evolution from GSM to UMTS

Evolution is one of the most common terms used in the context of UMTS. Generally it is understood to mean the technical evolution, i.e. how and what kind of equipment and in which order they are brought to the existing network if any. This is partly true but in order to understand the impact of the evolution, a broader context needs to be examined. Evolution as a high-level context covers not only the technical evolution of network elements but also expansions to network architecture and services. When these three evolution types are going hand in hand the smooth migration from 2G to 3G will be successful and generate revenue (Figure 2.1).

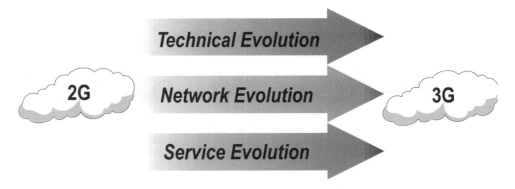

Figure 2.1. Evolution types

Technical evolution means the development path of how network elements will be implemented and with which technology. This is a very straightforward development and follows strictly the general, common technology development trends. Because network elements together form a network, the network will evolve accordingly, in theory. In this phase one should bear in mind that a network is as strong as its weakest element and due to the open interfaces defined in the specifications many networks are combinations having equipment provided by many vendors. Technical evolution may proceed however with different paces in association with different vendor's equipment and when adapting evolution-type changes between several vendors' equipment the result may not be as good as expected.

Service evolution is not such a straightforward issue. It is based on demands generated by the end-users and these demands could be real or imagined; sometimes network operators and equipment manufacturers offer services way beyond subscriber expectations. If the end-users' needs and operators' service palette do not match each other, difficulties with cellular business can be expected.

The main idea behind the GSM specifications was to define several open interfaces, which determine the standardised components of the GSM system. Because of this interface openness, the operator maintaining the network may obtain different components of the network from different GSM network suppliers. Also, when an interface is open it defines strictly how system functions are proceeding at the interface and this in turn determines which functions are left to be implemented internally by the network elements on both sides of the interface.

As was experienced when operating analogue mobile networks, the centralised intelligence generated a lot of load in the system, thus decreasing the overall system performance. That is why the GSM specification in principle provided the means to distribute intelligence throughout the network. The above-mentioned interfaces are defined into places where their implementation is both natural and technically reasonable.

From the GSM network point of view, this decentralised intelligence is implemented by dividing the whole network into four separate subsystems being Network Subsystem (NSS), Base Station Subsystem (BSS), Network Management Subsystem (NMS) and Mobile Station (MS) (Figure 2.2).

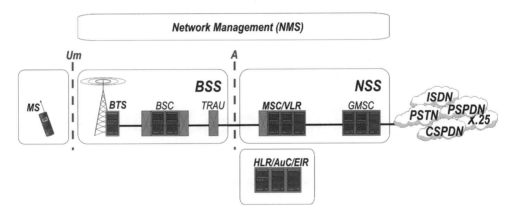

Figure 2.2.   Basic GSM network and its subsystems

The actual network needed for call establishing is composed of the NSS, the BSS and the MS. The BSS is a network part responsible for radio path control. Every call is connected through the BSS. The NSS is a network part taking care of call control functions. Every call is always connected by and through the NSS. The NMS is the operation and maintenance related part of the network. It is also needed for the whole network control. The network operator observes and maintains the quality and services of the network through the NMS. The open interfaces in this concept are located between the MS and BSS (Um interface) and between the BSS and NSS (A interface). The interface between the NMS and the NSS/BSS was expected to be open, but its specifications were not ready in time and this is why every manufacturer implements NMS interfaces with their own proprietary methods.

The MS is a combination of terminal equipment and a subscriber's service identity module. The terminal equipment as such is called Mobile Equipment (ME) and the subscriber's data is stored in a separate module called the Service Identity Module (SIM). Hence, ME + SIM = MS.

The Base Station Controller (BSC) is the central network element of the BSS and it controls the radio network. This means that the following functions are the BSC's main responsibility areas: maintaining radio connections towards the MS and terrestrial connections towards the NSS. The Base Transceiver Station (BTS) is a network element maintaining the air interface (Um interface). It takes care of air interface signalling, ciphering and speech processing. In this context, speech processing means all the methods BTS performs in order to guarantee an error-free connection between the MS and the BTS. The Transcoding and Rate Adaptation Unit (TRAU) is a BSS element taking care of speech transcoding, i.e. it is capable of converting speech from one digital coding format to another and vice versa.

The Mobile Services Switching Centre (MSC) is the main element of the NSS from the call control point of view. MSC is responsible for call control, BSS control functions, interworking functions, charging, statistics and interface signalling towards BSS and interfacing with the external networks (PSTN/ISDN/packet data networks). Functionally the MSC is split into two parts, though these parts could be in the same hardware. The serving MSC/VLR is the element maintaining the BSS connections, mobility management and interworking. The Gateway MSC (GMSC) is the element participating in mobility management, communication management and connections to the other networks. The Home Location Register (HLR) is the place where all the subscriber information is stored permanently. The HLR also provides a known, fixed location for the subscriber-specific routing information. The main functions of the HLR are subscriber data and service handling, statistics and mobility management. The Visitor Location Register (VLR) provides a *local* store for all the variables and functions needed to handle calls to and from mobile subscribers in the area related to the VLR. Subscriber related information remains in the VLR as long as the mobile subscriber visits the area. The main functions of the VLR are subscriber data and service handling and mobility management. The Authentication Centre (AuC) and Equipment Identity Register (EIR) are NSS network elements taking care of security-related issues. The AuC maintains subscriber identity-related security information together with the VLR. The EIR maintains mobile equipment identity (hardware) related security information together with the VLR.

When thinking of the services, the most remarkable difference between 1G and 2G is the presence of the data transfer possibility; basic GSM offers 9.6 kb/s symmetric data connection between the network and the terminal. The service palette of the basic GSM is directly adopted from Narrowband ISDN (N-ISDN) and then modified to be suitable for mobile network purposes. This idea is visible throughout the GSM implementation; for example, many message flows and interface handling are adapted copies of corresponding N-ISDN procedures.

The very natural step to develop the basic GSM was to add service nodes and service centres on top of the existing network infrastructure. The GSM specifications define some interfaces for this purpose, but the internal implementation of the service centres and nodes are not the subject of those specifications. The common name for these service centres and nodes is Value Added Service (VAS) Platforms and this term describes quite well the main point of adding these equipment to the network (Figure 2.3).

The minimum VAS platform contains typically two pieces of equipment; Short Message Service Centre (SMSC) and Voice Mail System (VMS). Technically speaking the VAS platform equipment is relatively simple and meant to provide a certain type of service. They use standard interfaces towards the GSM network and may or may not have external interfaces towards other network(s).

From the service evolution point of view, VAS is the very first step in generating revenue

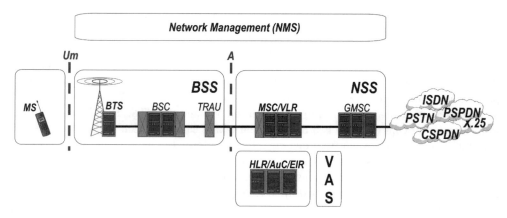

Figure 2.3. Value added service platform

with services and partially tailoring them. The great successor in this sense has been the SMS, which was originally planned to be a small add-in in the GSM system. Nowadays it has become extremely popular among GSM subscribers.

Basic GSM and VAS are basically intended to produce "mass services for mass people" but due to requirements raised from end-users, a more individual type of services is required. To make this possible, the Intelligent Network (IN) concept (Figure 2.4) was integrated together with the GSM network. Technically this means major changes in switching network elements in order to add the IN functionality and, in addition, the IN platform itself is a relatively complex entity. IN enables service evolution to take big steps towards individuality and also with IN the operator is able to perform more secure business, for example, pre-paid subscriptions are mostly implemented with IN technology.

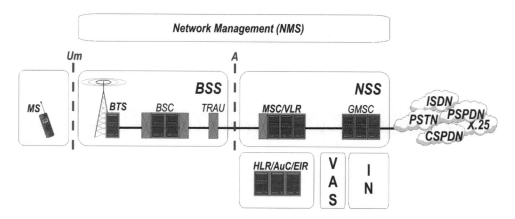

Figure 2.4. Intelligent network

In the beginning, GSM subscribers have used the 9.6 kb/s circuit switched symmetric "pipe" for data transfer. Due to the Internet and electronic messaging the pressures for mobile data transfer have increased a lot and this development was maybe underestimated at the time when the GSM system was specified. To ease this situation, a couple of enhancements have been introduced. Firstly, the channel coding is optimised. By doing this the effective bit rate has increased from 9.6 kb/s up to ≈14 kb/s. Secondly, to put more data through the air interface, several traffic channels can be used instead of one. This arrangement is called High Speed Circuit Switched Data (HSCSD) (Figure 2.5). In an optimal environment a HSCSD user may reach data transfer with 40–50 kb/s data rates. Technically this solution is quite straightforward but unfortunately it wastes resources and some end-users may not be happy with the pricing policy of this facility; the use of HSCSD very much depends on the price the operators set for its use. Another issue is the fact that most of the data traffic is asymmetric in nature, i.e. typically a very low data rate is used from the terminal to network direction (uplink) and higher data rates are used in the opposite direction (downlink).

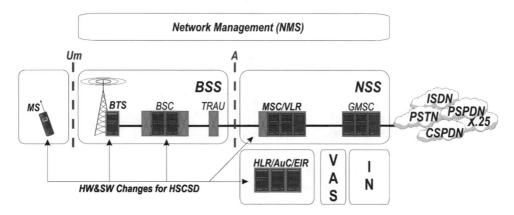

Figure 2.5.   Channel coding and HSCSD effects

The circuit switched symmetric Um interface is not the best possible access media for data connections. When also taking into account that the great majority of the data traffic is packet switched in nature, something more had to be done in order to "upgrade" the GSM network to make it more suitable for more effective data transfer. The way to do this is General Packet Radio Service (GPRS) (Figure 2.6). GPRS requires two additional mobile network specific service nodes: Serving GPRS Support Node (SGSN) and Gateway GPRS Support Node (GGSN). By using these nodes the MS is able to form a packet switched connection through the GSM network to an external packet data network (the Internet).

GPRS has the possibility to use asymmetric connections when required and thus the network resources are utilised better. GPRS is a step bringing IP mobility and the Internet *closer* to the cellular subscriber but it is not a complete IP mobility solution. From the service point of view, GPRS starts a development path where more and more traditional circuit switched services are converted to be used over GPRS because those services were originally more suitable for packet switched connections. One example of this is Wireless Application Protocol (WAP), the potential of which is to be discovered when using GPRS.

When packet switched connections are used, the Quality of Service (QoS) is a very essential issue. In principle the GPRS supports the QoS concept but in practise it does not. The reason here is that GPRS traffic is always second priority traffic in the GSM network: it

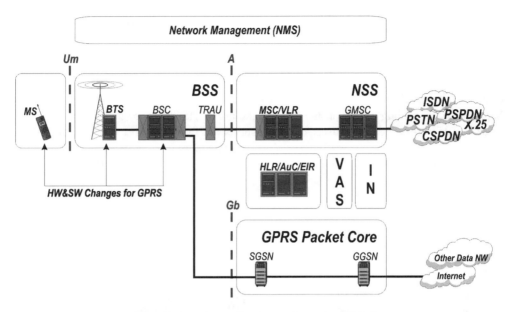

Figure 2.6.   General packet radio service (GPRS)

uses otherwise unused resources in the Um interface. Because the amount of unused resources is not exactly known in advance, no one can guarantee a certain bandwidth for the GPRS continuously and thus the QoS cannot be guaranteed either.

By applying a completely new air interface modulation technique, Octagonal Phase Shift Keying (8-PSK), where one air interface symbol carries a combination of three information bits, the bit rate in the air interface can be remarkably increased. When this is combined together with very sophisticated channel coding technique(s), one is able to achieve a data rate of 48 kb/s compared to conventional GSM which can carry 9.6 kb/s per channel and one information bit is one symbol in the air interface. These technical enhancements are called Enhanced Data Rates for Global/GSM Evolution (EDGE) (Figure 2.7).

Development of the EDGE concept is divided into two phases, EDGE Phase1 and EDGE Phase2. EDGE Phase1 is also known as E-GPRS (Enhanced GPRS). Also the BSS is renamed as E-RAN (EDGE Radio Access Network). EDGE Phase1 defines channel coding and modulation methods that enable up to 384 kb/s data rates for packet switched traffic under certain conditions. The assumption here is that one GPRS terminal gets eight air interface time slots for one connection, thus 8 × 48 kb/s = 384 kb/s. In addition, the EDGE terminal must be close to the BTS in order to use a higher channel coding rate. EDGE Phase2 contains guidelines on how this same speed is achieved for circuit switched services. EDGE Phase2 is also commercially known as E-HSCSD.

From a network evolution point of view, EDGE in general has its pros and cons. A good point is the data rate(s) achieved; these are almost equal to UMTS urban coverage requirements. The disadvantage with EDGE is that the data rates offered are not necessarily available throughout the cell. If EDGE is to be offered with complete coverage, the amount of cells will remarkably increase. In other words, EDGE may be expensive solution in some cases. The future of EDGE is still to be seen since it has to compete with the true 3G solutions.

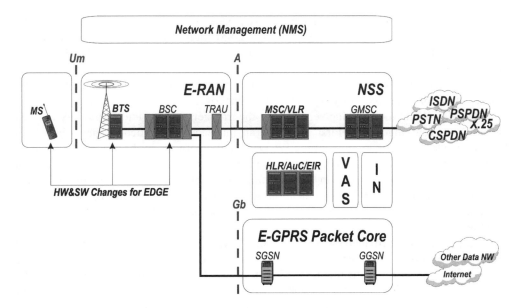

Figure 2.7.   Enhanced data rates for global evolution

Figure 2.8 represents a scenario how a 3G network is implemented according to 3GPP R99 specifications.

3G introduces the new radio access method, WCDMA. WCDMA and its variants are global, hence all 3G networks should be able to accept access by any 3G network subscriber. In addition to globality, WCDMA has been thoroughly studied in laboratory premises and it has been realised that it has better spectral efficiency than TDMA (in certain conditions) and it

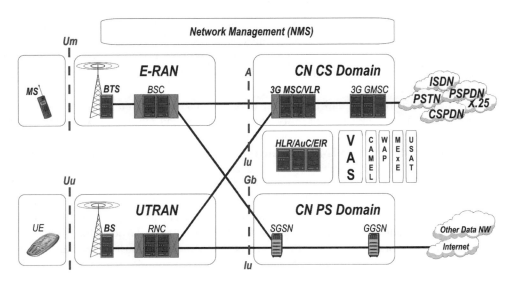

Figure 2.8.   3G network scenario (3GPP R99)

is more suitable for packet transfer than TDMA based radio access. WCDMA and radio access equipment as such are not compatible with GSM equipment, and this is why adding the WCDMA to the network one must add new elements: Radio Network Controller (RNC) and Base Station (BS).

On the other hand, one of the key requirements for UMTS is GSM/UMTS interoperability. One example of interoperability is inter-system handover, where the radio access changes from GSM to WCDMA and vice versa during the transaction. This interoperability is taken care of by two arrangements. First, the GSM air interface is modified so that it is able to broadcast system information about the WCDMA radio network in the downlink direction. Naturally the WCDMA radio access network is able to broadcast system information about surrounding the GSM network in the downlink direction, too. Second, to minimise the implementation costs, the 3GPP specifications introduce possibilities to arrange inter working functionality with which the evolved 2G MSC/VLR becomes able to handle the wideband radio access, UTRAN.

So far, the abilities provided by the IN platform have been enough from the service point of view. The concept of IN is directly adopted from the PSTN/ISDN networks and thus it has some deficiencies as far as the mobile use is concerned. The major problem with standard IN is that the IN as such is not able to transfer service information between networks. In other words, if a subscriber uses IN based services they work well but only within his/her home network. This situation can be handled by using "evolved IN" called Customised Applications for Mobile network Enhanced Logic (CAMEL). CAMEL is able to transfer service information between networks. Later on, the role of CAMEL will increase a lot in 3G implementation; actually almost every transaction performed through the 3G network will experience CAMEL involvement at least to some extent.

Transmission connections within the WCDMA radio access network are implemented by using ATM on top of a physical transmission media (3GPP R99 implementation). A pre-standardisation project FRAMES (1996–1998) discussed a lot whether to use ATM in the network or not. The final conclusion was to use ATM because of two reasons:

- ATM cell size and its payload are relatively small. The advantage here is that the need of information buffering decreases. If buffering a lot, expected delays will easily increase and also the static load in the buffering equipment will increase. One should bear in mind that buffering and thus generated delays have a negative impact on the QoS requirements of real-time traffic.
- The other alternative, IP, and its version IPv4 was also considered but IPv4 has some serious drawbacks, being limited in its addressing space and missing QoS. On the other hand, ATM and its bit rate classes match very well with QoS requirements. This leads to the conclusion that where ATM and IP are combined (for packet traffic), IP is used on top of ATM. This solution combines the good points of both protocols: IP qualifies the connections with the other networks and ATM takes care of the connection quality and also routing. Due to IPv4 drawbacks a compromise has been made. Certain elements of the network use fixed IPv4 types of addresses but the real end-user traffic uses dynamically allocated IPv6 addresses, which are valid within the 3G network. To adapt the 3G network to the other networks in this case, the 3G IP backbone network must contain an IPv4 $\leftrightarrow$ IPv6 address conversion facility, because the external networks may not necessarily support IPv6.

The core network nodes are evolved technically, too. The CS domain elements are able to handle both 2G and 3G subscribers. This requires changes in MSC/VLR and HLR/AC/EIR. For example, security mechanisms during the connection set-up are different in 2G and 3G and now these CS domain elements must be able to handle both of them. The PS domain is actually an evolved GPRS system. Though the names of the elements here are the same as that in 2G, their functionality is not. The most remarkable changes concern the SGSN, whose functionality is very different from that in 2G. In 2G, the SGSN is mainly responsible for Mobility Management activities for a packet connection. In 3G, the Mobility Management entity is divided between the RNC and SGSN. This means that every cell change the subscriber does in UTRAN is not necessary visible to the PS domain, but RNC handles these situations.

The 3G network implemented according to 3GPP R99 offers the same services as that of GSMPhase2+. This is, all the same supplementary services are available, teleservices and bearer services have different implementation but this is not visible to the subscriber; a speech call is still a speech call, no matter whether it is done through a traffic channel (GSM) or by using 3G bandwidth. In addition to GSM, the 3G network in this phase may offer some other services not available in GSM, for example, video call could be one of those. In this phase the majority of services are moved/transferred/converted to PS domain whenever reasonable and applicable. WAP is one of those candidates, because the nature of the information transferred in WAP is packet switched. The PS domain is taken into effective use and one service branch containing a variety of different services will be location based services utilising the subscriber location mechanisms built into the 3G network. The new positioning methods and related network elements are introduced in Chapter 7.

The development steps after 3GPP R99 are somewhat unclear in the level of details, but some major trends are visible. The main trends in the following development steps are separation of connection, its control and services and, at the same time, the conversion of the network to be completely IP based. From the service evolution point of view, these development steps also recognise that multimedia services should be provided by the 3G network itself. Multimedia means a service where at least two media components are combined, for example, voice and picture.

These trends are big issues as such and this is why they are implemented in phases, the first phase being 3GPP R4. The 3GPP R4 implementation introduces separation of connection, its control and services for the CN CS domain.

In the CN CS domain actual user data flow goes through Media Gateways (MGW), which are elements maintaining the connection and performing switching functions when required. The whole process is controlled by a separate element evolved from MSC/VLR called MSC server. One MSC server can handle numerous MGWs and thus the CN CS domain is freely scalable; when one wishes to add control capacity, a MSC server is added. When one desires to add switching capacity, MGWs are added.

When this kind of network has been set up, the pace of technology development and specifications set the next limit. The more momentum the IPv6 gains, the more of the 3G network connections that can be converted to IPv6, too. This decreases the need of IPv4 ↔ IPv6 conversions. In this phase the traffic relationship between circuit and packet switched will remarkably change. The majority of the traffic is packet switched and also some traditionally circuit switched services, like for example speech, will become at least partially packet switched (VoIP, Voice Over IP). For example, a conventional GSM call is actually changed to a VoIP call in the MGW where the BSS is connected. There are many ways to implement VoIP calls but the new CN subsystem called IMS (IP Multimedia Subsystem) is added here since it will offer uniform methods to perform VoIP calls. In addition to this the IMS is used for IP based multimedia services. Naturally the BSS part of the network could be implemented using IP but the time schedules of this change are unclear. The role of CAMEL will change, too. Because many of the services using CAMEL are converted from the circuit switched side to the packet switched side of the network, the CAMEL will now have connections also to the PS domain elements. In addition to this the CAMEL will be connecting elements between the service platforms and the network. This issue is explained in more detail in Chapter 7.

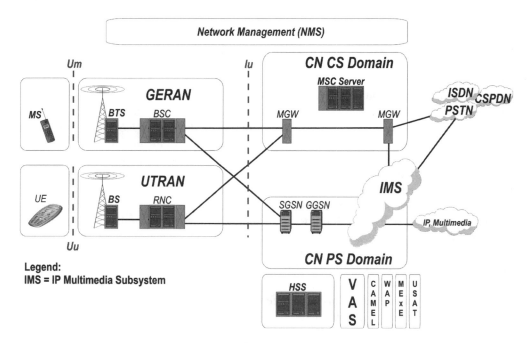

Figure 2.9.   3GPP R4 implementation scenario

In 3GPP R5 the evolution continues further and all traffic coming from UTRAN is supposed to be IP based. If we think of a voice call from UE to PSTN as an example, it is transported through UTRAN as packages and from the GGSN the VoIP call is routed to the PSTN via IMS, which provides required conversion functions.

From the UE point of view the network always "looks" the same in the development phases illustrated in Figures 2.8–2.10. Inside the network almost everything changes. The major change will happen in transport technology, which in 3GPP R99 implementation is ATM. The 3GPP R4 and R5 implementation scenarios aim to swap ATM for IP. Because the system must be backward compatible, the operator always has a choice whether to use ATM or IP as the transport technology or whether the solution contain both technologies. As explained earlier, the strength of ATM is its support for QoS at least in the beginning. As time goes by, the IP as a technology will contain QoS mechanisms implemented over various kinds of subnetworks, not only ATM.

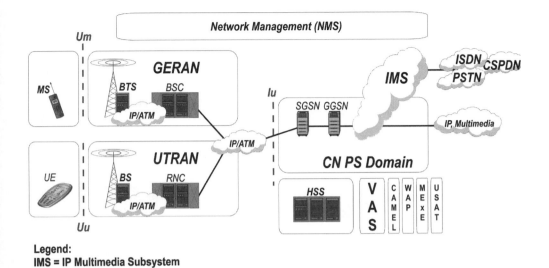

Figure 2.10.   Vision of 3GPP R5 (All IP)

In this phase the services and use of the network are more important than technology itself and due to this the used radio access technology may become less important. The main selection criterion for the used radio access technology is to offer enough bandwidth for the service used. The future vision here is that the 3G core network has interfaces for several radio access technologies, for instance, GSM, EDGE, CDMA2000, WCDMA and Wireless Local Area Network (WLAN). Naturally this sets many requirements for terminal manufacturers and the terminals capable of handling different kinds of access technology combinations will be introduced according to the market needs. 3G is now a fixed part of life and the network offers services and connections which have traditionally used other media. The 3G terminal becomes a combination of personal belongings, like a phone, purse, ID card and passport. In other words, it is very difficult to state where the computer ends and phone starts.

3

Basics of UMTS Radio Communication

The purpose of this chapter is to introduce briefly two fundamental radio communication issues to provide the reader, first, the reasons behind those restrictions and also opportunities when considering the basic architecture of any radio communication system. The second issue is related to how the radio communication is specifically handled in UMTS. This is addressed by describing the essentials of the third generation (3G) radio path and more specifically WCDMA-FDD, providing substantial mechanisms and terminology. More insight into the 3G radio path characteristics is given in Holma and Toskala (2001).

3.1. Radio Communication Fundamentals

Communication has always been the essential part of every kind of society and especially human society. Because social evolution has been an inherent characteristic of collective life resulting in more sophisticated interaction relationships, more advanced means have always been needed to meet the communication demands between members of society. During the history of human life, many techniques have been developed for the purpose of communication. Among them, radio communication has been, and will be one of the most important techniques that man has ever used for communicating.

The usage of radio communication has been realised since Hertz experimentally showed the relationship between light and electricity in 1887 after Maxwell had illustrated the principal equations of electromagnetic fields in 1864. Then, Marconi used the "Hertzian" wave to communicate, inventing wireless telegraphy in 1896. Exploiting the radio wave for transmitting information is the basic part of every kind of radio communication.

The fundamental principle of radio communication is that it utilises radio waves as a transmission medium. As natural phenomena, radio waves are a consequence of electromagnetic fields. Under certain circumstances, time dependent electromagnetic fields produce waves that radiate from the source to the environment. This source can be for example a transmitter like a base station or mobile handset. Because radio waves in general are based on electromagnetic fields, their characteristics are strictly dependent on the environment where the waves propagate. As a consequence, a radio wave based radio communication system is vulnerable to the environmental factors, for instance, mountains, hills, huge reflectors like buildings, the atmosphere, and so on.

Every communication system consists of at least two elements, which are the transmitter and the receiver. As it is the case in mobile systems, these two elements can be integrated in one device (transceiver) so that it is capable of operating both as a transmitter and receiver device. An example of such a device is the base station and mobile handset in any advanced public mobile system. Figure 3.1 illustrates the simplest radio communication system, consisting of one base station and one mobile handset. Suppose that the base station acts as a transmitter source for a specific time with certain environmental circumstances. Then the radio signal propagates from base station to the mobile station and with the speed of light. The received signal strength at the handset depends on the distance from the base station, the wavelength, and the communication environment.

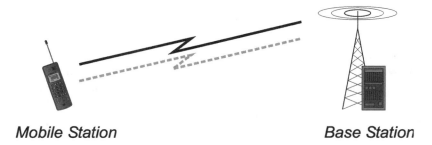

Mobile Station   **Base Station**

Figure 3.1.   Essential elements of a radio communication system

A radio wave propagation mechanism closely depends on the wavelengths or frequencies. In addition to that, any man-made or natural obstacle like high buildings, terrain, weather condition, etc. affects the way and time of signal propagation between the transmitter and receiver. Similarly, system parameters, for example antenna height and beam direction have naturally their own effect on the propagation distance, mode, and delay. The nature of radio communications inherently brings about some thorny problems. The main problems that every radio communication faces are as follows:

- Multipath propagation phenomena
- Fading phenomena
- Radio resource scarcity

Multipath propagation is also considered, by many means, as an advantage of the radio communication because it enables the radio receiver to hear the base station even without signal Line Of Sight (LOS). Despite that, it brings complexity to the system by setting specific requirements and constraints for it. In order to understand the nature of a radio communication system those characteristics should be well understood. Therefore, we explain them here in more detail.

The factors that effect radio propagation are extremely dynamic, sophisticated, and diverse. In spite of that and in order to model the propagation phenomena, the mechanisms can be classified into reflection, diffraction, and scattering phenomena (Figure 3.2). In the radio network environment these propagation mechanisms lead to multipath propagation, which causes fluctuations in the received signal's characteristics, including amplitude, phase, and angle of arrival, giving rise to multipath fading. Reflection is the consequence of collision

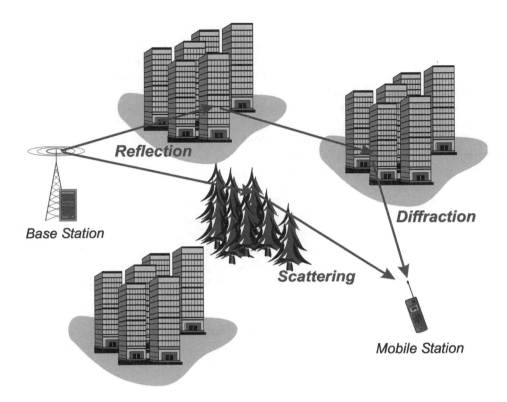

Figure 3.2. Propagation mechanisms: reflection, scattering and diffraction

of the electromagnetic wave with an obstruction whose dimensions are very large in comparison with the wavelength of the radio wave. The result of this phenomenon is reflected radio waves, which can be captured constructively at the receiver, e.g. mobile or base station. Diffraction, also called shadowing, in turn, is the consequence of collision of the radio wave with a obstruction which is impossible to penetrate. Scattering on the other hand is the consequence of the collision of the radio wave with obstructions whose dimensions are almost equal to or less than the wavelength of the radio wave. These phenomena together explain how radio waves can travel in a radio network environment without an LOS path.

From the receiver's perspective and depending on the existing preconditions of any above-mentioned propagation mechanisms, the received signal power is affected randomly by each or combination of these mechanisms. In addition to those propagation mechanisms, in mobile system the concepts of mobility, indoor coverage, outdoor coverage, and hierarchical network structure raise some specific aspects to the propagation environment, which makes the situation more sophisticated.

There are mainly two different ways to describe the effect of the propagation mechanism on the signal strength of the radio channel, including link budget and time dispersion. The basic idea behind the link budget is to determine the expected signal level at a particular distance or location from a transmitter like base station or mobile station. By modelling the link budget essential parameters of the radio network such as transmitter power requirements, coverage area, and battery life can be defined.

Link budget calculation, can be carried out by estimating the signal path-loss. Path-loss estimation can be done based on the free space model, defining that in an idealised free space model the attenuation of signal strength between the base station and mobile station likely behaves according to an inverse-square law. Due to radio environment differences in advanced mobile systems, it is almost impossible to consider all parameters affecting the radio channel modelling and system designing. Therefore, some general models, for the most usual cases have been developed (Okumura et al.). In order to consider the entire effect of radio channel fading, however, link budget alone cannot be adequate. In addition to that the effect of multipath propagation in terms of time dispersion should also be considered. This can be done by estimating the different propagation delays related to the replicas of the transmitted signal which reaches the receiver.

A typical fading process is illustrated in Figure 3.3. As shown, any fading process has the curves of two different simple shapes, that is, the downward deep fades, referring to the deterioration of the signal strength and the upraise curves, causing the undesired interference. Therefore, a combination of these simple curves can approximate the envelope of any fading process and the control actions for the fading process are separately determined to compensate for the round deviation of the signal level around the desirable average.

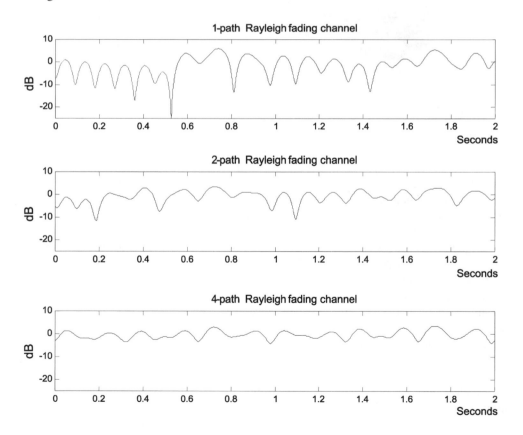

Figure 3.3.   A typical fading power signal

Figure 3.4 illustrates the main fading types, which may arise in any radio network environment in one way or another depending on the radio network environment. The main classes of the radio channel fading include large-scale and small-scale fading. The former represents the average signal power attenuation or path loss due to the mobile's motion over areas between the base station and the mobile station. Small-case fading, on the other hand, is the result of rapid variations in signal amplitude and phase that can be experienced between the base station and mobile station. Small-scale fading is also called Rayleigh fading or Rician fading depending on the NLOS or LOS characteristic of the reflective paths. Small-scale fading can be divided into frequency selective fading, flat fading, fast fading, and slow fading.

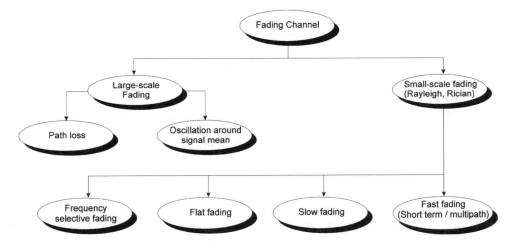

Figure 3.4.   Radio channel fading classification

The radio channel experiences signal fading that most likely consists of the Rayleigh faded component without LOS, making the channel more severe to cope with. As a result, a signal arrives at a mobile station from many directions with different delays, causing significant phase differences in signals travelling on different branches. Moreover, the behaviour of the radio channel fading is strictly dependent on the physical position and motion of the mobile station. A small change in position or motion of the mobile station may result in a different phase for each signal branch. Thus the motion of the mobile station through the radio network may result in a quite rapid fading phenomenon, which can be extremely disruptive to the radio signals, requesting considerable requirements from the radio network optimisation.

If we assume that a radio signal is transmitted from the base station to the mobile station with constant transmitted signal strength, then, the effect of fading can be seen as a function of distance of the strength of the base station transmitted signal as received at a mobile station. Therefore, as distance increases between the mobile station and base station, the received signal strength decreases at the mobile station. The average path loss – which basically follows log-normal distribution of average path loss – is the same for both directions.

Multiple path signal propagation causes rapid fading as well. Indeed, this type of fading is more difficult to compensate for because the signal arrives from these multiple paths in random phases and amplitude, resulting in Rayleigh fading characteristics. If different bandwidths are used in uplink and downlink in cellular systems then the Rayleigh fading is typically independent for downlink and uplink, meaning that when the uplink channel is fading, the downlink channel is not necessarily fading at the same time and vice versa.

In order for the mobile to adjust its transmitter power to correspond to the link path signal strength of its home cell base station it should take the inverse pace as the signal the base station experiences. Therefore, the mobile has to power up its signal in two control steps, that is, firstly it should compensate the log-normal fading and secondly fast fading. Sudden increase in transmitter power, however, may have drawbacks for the overall system performance in terms of capacity and service quality. As mobility is an inherent part of the mobile system, it causes more complexity to the radio channel fading. The duration of the fades from the median signal level is dependent on mobile station speed and the operating frequency in such a way that as the mobile station accelerates the fading becomes more severe.

In addition to fading and multipath phenomena, interference is a thorny problem for every radio communication system. The basic reason behind interference is that there are many simultaneous radio accesses to the base station. This is the case especially when the radio system uses a common share bandwidth. Moreover, many man-made noises may generate additional interferences to the radio system. However, the most dominant interference modes are basically caused by the common use of radio resources simultaneously. Minimising the undesirable impact of fading, interference and radio resource scarcity is considerably dependent upon the radio network planning approach, the utilised radio access techniques, and those algorithms used for controlling the radio resources. Therefore, before describing further details of the interference aspects of a radio system, we proceed next by introducing the concept of a cellular system as a basic solution for capacity limitation of radio systems due to path-loss, fading and interference.

In order to solve the previously mentioned problems many solutions have been developed especially during last decades since radio communications have been used for public communications. An increase in the number of radio communication users together with simultaneous demand for large area network coverage and diverse services, however, set extremely tight requirements for radio communication systems. As a result, many advanced solutions have been utilised by public radio communication systems, e.g. cellular concepts, advanced radio resource allocation techniques, modulation techniques, advanced antenna techniques, and so on.

3.2. Cellular Radio Communication Principles

The simple radio communication system illustrated earlier in this chapter is unable to provide access service to the large number of end-users, that is, it faces capacity limitation. Let's outline the basic problems that the inherent characteristics of radio communications described earlier cause for such a simple system.

Firstly, the public radio communications should offer duplex communication for providing simultaneous two way communication. Therefore, it is not practical to offer such services by, for instance broadcasting the information. In addition to that, the signal strength at the receiver deteriorates as the distance between the transmitter and receiver increases, resulting in unacceptable QoS in the area far away from the transmitter. Secondly, every transmitter is capable of offering a limited number of radio links/channels to the end-users who intend to have a call simultaneously, resulting in capacity limitation. Therefore, the spectrum should be used more efficiently to meet the demanded radio access.

The target of cellular concept is to address these problems. The main idea is simple. Suppose we are planning a radio network for a large city where there are millions of mobile users. Based on the cellular concept, the large area is divided into a number of sub-areas called *cells*. Each cell has its own base station, which is able to provide a radio link for specific numbers of simultaneous users by emitting a low-level transmitted signal. Figure 3.5 illustrates an example of a seven cell cluster of a cellular network.

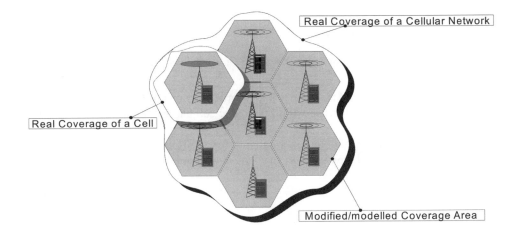

Figure 3.5.   A cluster of cells in a cellular network

The nature of radio communication and cellular concept are the primary reasons why the basic architecture of advanced mobile systems is currently what it is. Figure 3.6 illustrates the basic architecture of a cellular concept-based radio system. As shown, the basic architecture of any advanced cellular system consists of base stations, a switching network, and fixed network functionality for backbone transmission.

The cellular concept solution resolves the basic problems of radio systems in terms of radio system capacity constraints, but at the same time it encounters other problems, such as:

- Interference due to the cellular structure, including both inter- and intra-cell interference
- Problems due to mobility
- Cell-based radio resource scarcity

Assume a cellular system with asynchronous users sharing the same radio bandwidth and using the same radio base station in each coverage area or cell, each base station not only receives interference from mobiles in the home cell but also from terminals and base stations located in neighbouring cells. Therefore, depending upon on the source of interference, they can be classified as intra-cell/co-channel, inter-cell/adjacent cell, and interference due to thermal noise.

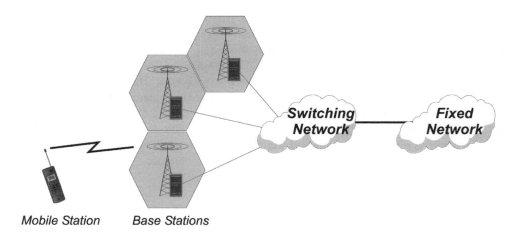

Mobile Station   Base Stations

Figure 3.6.   Basic structure of a cellular system

Therefore, and as illustrated in Figure 3.7, in order to cope with the total interference received from terminals, interference within the home cell and neighbouring cells should be investigated.

From the mobile station standpoint, inter-cell interference is caused from the signal of base stations or mobile stations of the neighbouring cells of the home cell where the mobile station is camping.

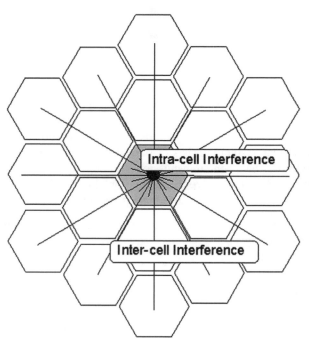

Figure 3.7.   Inter-cell and intra-cell interference in a cellular system

In a cellular system, if the same frequency is used in every cell then the mobile station, which is connected to the radio network, encounters the problem of channel interference, including adjacent channel interference and co-channel interference. In addition to that, the network environment may include mountains, hills, and other obstacles, causing multipath problems and path losses. This is solved mainly by utilising the *frequency reuse technique*, i.e. in each cell of cluster pattern a different frequency is used. So, the frequency reuse factor is an essential parameter for radio network planning in any cellular system, which operates based on frequency sharing principles. By optimising the frequency reuse factor the problems of adjacent channel and co-channel hazards can be removed considerably, resulting in increased capacity, which is crucial to the public radio systems. Figure 3.8 shows an example of one-seven and one-one frequency reuse patterns, respectively.

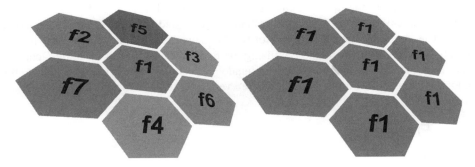

Figure 3.8.   Frequency re-use pattern, FR = 7 and FR = 1

The cellular concept increases the radio system capacity especially when utilising with frequency reuse. The smaller the cells, the more efficiently the radio spectrum is used but the cost of the system increases at the same time because more base stations are needed.

Multilayer network designing, including macro-, micro- and pico-cells is a further step in providing advanced system designing solutions in conjunction with the frequency reuse concept to improve the system capacity. It is the subject of network planning to optimise the combination of network structure and frequency reuse to increase the system capacity cost-effectively, avoiding undesirable increase in the number of needed base stations. These solutions, however, set new requirements for the system to handle the mobile's mobility and desirable system performance in terms of quality of services and signalling load.

Although, mobility allows the possibility of being reachable anywhere and any time for the end-user, nevertheless, it sets very strict requirements for the cellular system to support such a feature. Managing the mobile terminals' mobility is one of the most essential parts of any cellular system functionality. Generally, in a radio communication system, paging, location updating, and handover operations provide the user mobility. Handover mechanism guarantees that whenever the mobile is moving from one base station area/cell to another, the radio signal is handed over to the target base station. In addition, location update and paging mechanisms guarantee that the mobile station can be reached even though there is no continuous active radio link between the mobile and corresponding base station. The network always initiates the paging mechanism; mobile station initialises the location update procedure.

3.3. Multiple Access Techniques

The cellular concept approaches the capacity limitation in terms of system coverage. Therefore, it does not alone help the per-cell capacity limitation as far as the simultaneous users are in question. From the radio spectrum standpoint, it is extremely important how the radio resources are allocated to the simultaneous users. Controlling the radio resources has become one of the most critical features of any public radio communication system, which provides services to the huge numbers of service requesters.

Multiple accesses techniques have been developed to combat the problem of simultaneous radio access allocation to the access requesters. The main aspect of any multiple access scheme is how the available frequency band is allocated. The primary problems encountered by multiple access solutions are related to the intrinsic characteristics of radio systems; bandwidth limitation, multipath fading of the radio link and interference from other users in the cellular environment where the cellular concept and frequency reusing have been utilised. The key issue for effective use of the frequency is multiple access. That is, there should be as many users as possible utilising the fixed resource, in this case, frequency.

The multiple access method used in analogue cellular networks is called FDMA (Frequency Division Multiple Access) where every user uses their own frequency (or pair of frequencies) (Figure 3.9).

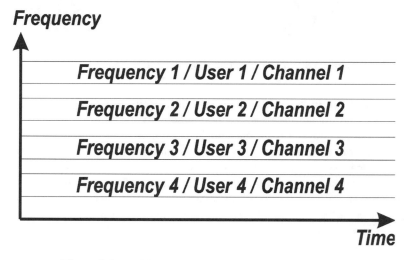

Figure 3.9.   FDMA (frequency division multiple access)

In the case of FDMA, having its own frequency or bandwidth specifies the radio channel. When considering the number of radio connections this means one frequency = one user = one channel. This multiple access technique, however, is not an efficient way of sharing the limited radio resources in a public mobile system. Therefore, a more effective way to utilise frequency resources which enables an increase in the system capacity, called TDMA (Time Division Multiple Access) was developed at the earlier phase of radio communication systems. TDMA is the most common multiple access method used in the second generation cellular systems such as in GSM. Also, TDMA based transmission systems are widely used. Maybe the most known system is the G.703 specification based PCM (Pulse Code Modulation) transmission.

In the GSM system Um interface the users of a given frequency divide it in time: every user has a little slice of time (timeslot) for different operations. This timeslot is repeated frequently and this generates an impression of continuous connection. This system makes it possible to have several users – as many as the number of timeslots, which is eight for GSM – simultaneously using the same frequency (Figure 3.10).

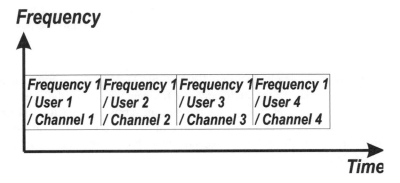

Figure 3.10.   TDMA (time division multiple access)

Code Division Multiple Access (CDMA) is another multiple access technique which has been used for similar purposes as FDMA and TDMA. However, it approaches the same problem by utilising a totally different strategy. As a spread-spectrum-based radio access scheme, CDMA is one of the most sophisticated ones, which has been used in different applications. We present a short review of the basic CDMA scheme without going into the subject in detail. Figure 3.11 illustrates the basic radio resource strategy of the CDMA scheme. Unlike in TDMA and FDMA schemes, the radio resource is allocated based on codes in the CDMA scheme. Thus all the simultaneous users can occupy the same bandwidth

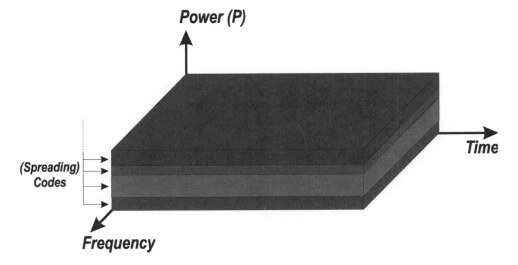

Figure 3.11.   CDMA (code division multiple access)

at the same time. Every user is assigned a code/codes varying per transaction and those codes are used for cell, channel and user separation. Every user uses the same frequency band simultaneously and hence there is no "time slots" or frequency allocation in the same sense as in TDMA and FDMA based systems, respectively.

- If the originating bit rate is low, it can be spread well and thus the power required for transmission will be small. This kind of case can be seen as a thin layer in Figure 3.11.
- If the originating bit rate is high it cannot be spread as well and thus the power required for transmission will be higher. This kind of case can be seen as a thick layer in Figure 3.11.

Unlike bandwidth-limited multiple access (FDMA and TDMA) which suffers mainly from co-channel interference due to the high range of frequency reusing, the inter-user uplink interference is the most crucial type of interference in terms of capacity and QoS in the CDMA system. The reason is that inter-user uplink interference increases on a power basis, and the performance of each user becomes poorer as the number of simultaneous users increases in a single cell.

Depending on the spreading signal used in the modulation, the CDMA scheme can be categorised into the following groups:

- Direct Sequence CDMA (DS-CDMA)
- Frequency Hopping CDMA (FH-CDMA)
- Time Hopping CDMA (TH-CDMA)
- Hybrid Modulation CDMA (HM-CDMA)
- MultiCarrier CDMA (MC-CDMA)

In a DS-CDMA scheme, the data signal is scrambled by the user-specific pseudo noise (PN) code at the transmitter side, for example mobile or base station to make the spreading of the signal with the desirable chip rate and process gain. At the receiver (mobile or base station), the original signal is extracted by exactly the same spreading code sequence. As a result, every signal is assumed to spread over the entire bandwidth of the radio connection. Interference may therefore be generated from all directions in contrast to narrow-band cellular systems (Figure 3.12).

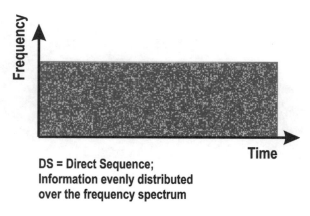

**DS = Direct Sequence;
Information evenly distributed
over the frequency spectrum**

Figure 3.12. DS-CDMA

The frequency reuse for DS-CDMA is one, meaning that all users have the same frequency band to carry their information. The inherent advantage of the DS-CDMA is its multipath fading tolerability. Indeed, in a TDMA-based system like GSM the similar advantage is achieved by utilising a frequency-hopping technique, resembling the spread spectrum scheme. On the other hand, in the case of common shared frequency it is vulnerable to multiuser interference more than TDMA and FDMA.

In FH-CDMA, changing the carrier frequency over the transmitted time of the signal produces the spread bandwidth. That is, during a specific time span, the carrier frequency is kept the same, but after that, the carrier is hopped to another frequency based on the spreading code, resulting in a spread signal (Figure 3.13).

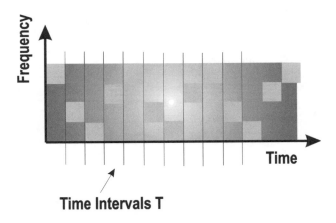

Figure 3.13.   FH-CDMA

Depending on the hopping rate of the carrier signal, FH-CDMA is divided into two parts: fast frequency hopping and slow frequency hopping. For the fast frequency hopping modulation scheme, the hopping rate is typically greater than the symbol rate and for the slow frequency scheme, the hopping rate is typically smaller than the symbol rate. Some essential procedures of the CDMA scheme, like power control, are much easier to implement and handle in FH-CDMA than in DS-CDMA. This is partly due to frequency division aspect of this technique when applying to the cellular systems. On the other hand, producing a high-rate frequency hopping is, however, a complex issue in FH-CDMA.

In the time hopping CDMA scheme, the transmitted signal is divided into frames, and those frames are further divided into time intervals. During the transmission time, the data burst is hopped between frames basically by utilising code sequences. Principally, the aim is to choose the code sequence so that the simultaneous signal transmission in the same frequency may be minimised. The most important idea behind this approach is to create synergy between the strengths of the previously described schemes to provide convenient schemes for specific applications. The combinations of these CDMA schemes leads to different hybrid schemes such as DS/FH, DS/TH, FH/TH, and DS/FH/TH.

Unlike DS-CDMA, in MC-CDMA the entire frequency band is used with several carriers instead of one. Therefore, the MC-CDMA transmitter basically spreads the original data over different sub-carriers using different frequency bands by denoting a spreading code in the frequency demand (Figure 3.14). The MC-CDMA schemes are mainly categorised into two groups. One spreads the original data stream using a given spreading code, and then modulates different sub-carriers. The other one is basically similar to DS-CDMA.

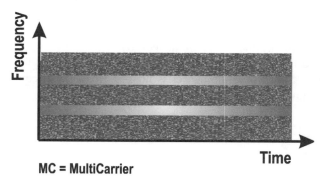

MC = MultiCarrier

Figure 3.14. MC-CDMA

3.4. Regulation

Conventionally, regulation has been an inseparable part of the development in communication systems. The role of regulation is concretised in two main areas, radio spectrum allocation and overall technical standardisation of the communication systems. In both areas, the main drivers of the regulation rely on an technology–engineering basis and economic–political aspects. Legal administration plays an important role in supporting the entire regulation process (Figure 3.15).

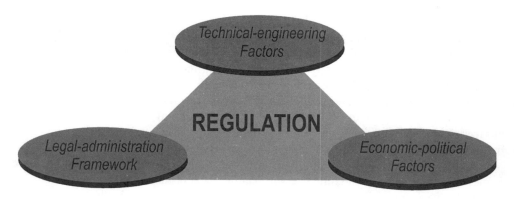

Figure 3.15. Interaction of different factors in the regulation process

So far, we have described the different techniques applied in radio systems to utilise the radio spectrum as efficiently as possible. Unlike the public understanding, the radio spectrum is not an unlimited natural resource but rather it is extremely scarce. Although, the advanced systems exploit the available radio spectrum more efficiently, nevertheless, rapid growth in demand for wireless services increases the scarcity of the spectrum. Therefore, it is essential to regulate the use of the radio spectrum to avoid non-harmonised and inefficient use of such a scarce natural resource. Indeed, radio spectrum regulation has always been a cornerstone of the communication business. In order to understand the role of radio spectrum regulation two main aspects should be distinguished, technical–engineering and economic–political aspects. Regarding the technical aspect of the radio spectrum, spectral efficiency is of primary concern in designing of any wireless system. As we mentioned earlier in this chapter, many technical approaches, i.e. multiple accesses, cellular concept, etc. have been developed to use the radio spectrum as efficient as possible.

Figure 3.16 illustrates the electromagnetic spectrum, including the radio wave spectrum from 3 kHz up to 300 GHz. On the other hand, the figure shows that most of the allocated radio spectrum allocated for wireless applications ranges from 100 to over 2000 MHz. The reason is partly because of the characteristics of the radio wave, that is, the higher the radio frequency the higher the radio wave attenuation. Therefore, technically, the higher the radio frequency the system implementation is more complex and non-cost-effective. In a cellular system, higher frequency application may increase the number of base stations needed in the subsystem for providing the required coverage.

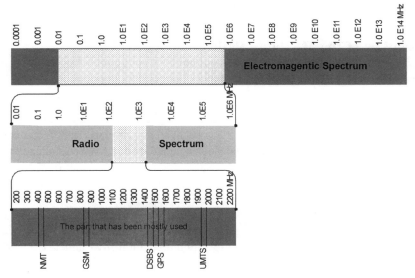

Figure 3.16.   The electromagnetic spectrum, radio wave portion and the most demanded portion

The economic–political aspects of the radio spectrum are still more complicated. The general national radio administrations are in charge of careful regulation of the whole radio spectrum. In addition, many international organisations are involved in radio spectrum regulation. The basic goal is to make the scarce radio resource available, in a manner that, on one hand, the system suppliers and users are content, and on the other hand, the spectrum has been used efficiently.

In addition to these factors, the communication market environment is extremely competition-oriented and continuously changing. Thus, spectrum regulation has also become a key competitive factor both nationally and globally and hence the process of deregulation is bringing about totally new figures to the field. Moreover, national interests are not always aligned, resulting in divergent spectrum allocation. For instance, a radio spectrum allocated to military systems has been a high priority in conjunction with international joint spectrum allocation strategy.

Due to dramatic growth in the wireless communication market, it seems that the scarcity of the radio spectrum will be more crucial in the future. Therefore, it is essential to make a trade-off between technical solutions, wireless applications, and the regulation to obtain a fair level of compromise solution. In terms of 1G, the radio spectrum was something to be given away free-of-charge except in the US, where radio spectrum slots have always been subject to regulation fees. The free-of-charge radio spectrum principle was going to be continued in 2G and this development was partially successful.

The spectrum regulation for the 3G system is bringing more competition to the field of communication technology and business. It is also bringing a common understanding that the radio frequency is not an endless natural resource as the number of mobile users is dramatically increasing and this technology is converging with the Internet. The radio spectrum regulation fees increased to a very high level. 3G spectrum regulation aspects are a complete story of their own and estimating their consequences is not within the scope of this book.

3.5. Essentials of the 3G Radio Path

3.5.1. Frequency Band and Regulatory Issues

3G spectrum regulation has been a hot subject of discussion during the last few years worldwide. It includes many sophisticated issues, which vary greatly between countries and operators. Firstly, the needed spectrum room for 3G is not available similarly in different countries. That is basically because of the spectrum portion occupied by other systems already in use or the guard bands needed for the systems, which operate on adjacent frequency bands. Secondly, the political approaches for spectrum licensing also vary from country to country. Naturally, operators involved in 2G systems are willing to have the edge technology by participating in 3G development work. In addition, many Greenfield operators (those newcomers to the mobile communications business) intend to reinforce their rule in the market by launching the 3G system. Therefore, the 3G spectrum regulation environment demands close international and national co-operation in order to harmonise spectrum allocation world-wide and manage it efficiently within countries (Figure 3.17).

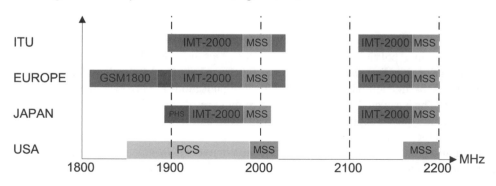

Figure 3.17.   The 3G system spectrum

From a 3G point of view, it has been decided (in mid-1999 by OHG) that there will be three CDMA variants in use. These are:

- DS-WCDMA-FDD: Direct Sequence – Wideband Code Division Multiple Access – Frequency Division Duplex
- DS-WCDMA-TDD: Direct Sequence – Wideband Code Division Multiple Access – Time Division Duplex
- MC-CDMA: Multi Carrier – Code Division Multiple Access

In these terms, the first part describes the information spreading method over the frequency spectrum, the second part indicates multiple access scheme and the last part expresses how the different transmission directions (uplink and downlink) are separated. Before proceeding it should be noted that the word "Wideband" here does not have any specific meaning. Originally it was inserted to the name because the Euro-Japanese version of CDMA used wider bandwidth than the American one. The wider band makes it possible to insert some attractive features to the system like, for instance, multimedia services with adequate bandwidth and macro diversity. Due to the OHG decision, however, all the three CDMA variants use similar bandwidth.

The WCDMA-FDD uses frequencies of 2110–2170 MHz downlink (from the BS to the UE) and 1920–1980 MHz uplink (from the UE to the BS). Air interface transmission directions are separated by different frequencies and duplex distance is 190 MHz.

The TDD variant of the WCDMA uses a frequency band located in both sides of the WCDMA-FDD uplink. The lower frequency band offered for the TDD variant is 20 MHz and the higher one is 15 MHz.

For the purpose of comparison it should be mentioned that the GSM1800 system uses frequencies of 1805–1880 MHz downlink (from the BTS to the MS) and 1710–1785 MHz uplink (from the MS to the BTS). Air interface transmission directions are separated from each other by different frequencies and duplex distance is 95 MHz.

In this book we will shortly introduce the DS-WCDMA-FDD basics and related technology, since it is the first radio access technology to be used when implementing UMTS wideband radio access. Due to simplicity, we further use the name "WCDMA" as a synonym of "DS-WCDMA-FDD".

3.5.2. Basic Concepts

The principles of the WCDMA technique are based on spread spectrum. Therefore, in order to outline the WCDMA scheme, it is essential to understand the general principles of spread spectrum. We next describe briefly the highlights of the spread spectrum scheme.

Spread spectrum is a well-proven modulation scheme that creates a bandwidth for the transmitted signal much wider than the bandwidth of the actual information, which is intended for transmission. The history of the spread spectrum technique goes back almost to the 1950s since the use of the scheme for military communication become a reality. Thanks to the efficient use of the radio spectrum by allowing additional users to share the same frequency and utilising efficient mechanisms like fast power control, diversity and soft handover the CDMA variant of the spread spectrum scheme has become a very promising radio access alternative for public applications of mobile communication systems. The main advantages of the spread spectrum scheme can be summarised as follows:

- Its resistance to radio interference and jamming
- It lowers the probability of interception by an adversary
- It is resistant to signal interference from multiple transmission signal branches
- It provides multiple access facility with a reuse factor equal to one
- It supports the means for measuring range, or the distance between two points
- It yields the possibility of utilising diversity techniques, including multipath diversity, as well as frequency and time diversity.
- It provides user access at any time without waiting for a free channel as far as the level of interference meets the system's tolerance.

Having outlined the main characteristics of the spread spectrum scheme, we further present the basic modulation technique of the scheme. Figure 3.18 illustrates the main principles of the modulation process of the DS spread spectrum scheme. Assume that the radio signal is transmitted from the base station to the mobile station. At the base station, the transmitted signal with rate R is spread by convoluting with a wideband spreading signal, creating a spread signal with bandwidth W. At the mobile, the received signal is multiplied by exactly the same spreading signal. Now if the spreading signal, locally generated at the mobile, is synchronised with the spread code/signal, the result is the original signal plus possibly some trumped up higher frequency components which are not part of the original signal, and hence can easily be filtered. If there is any undesired signal at the mobile, on the other hand, the spreading signal will affect it just as it did the original signal at the base station, spreading it to the bandwidth of the spreading signal.

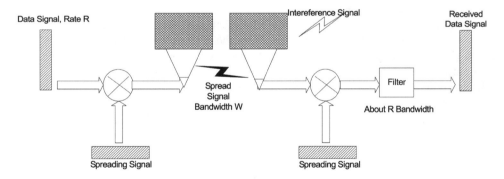

Figure 3.18. The basic technique of DS spread spectrum

In WCDMA, the data stream of the base station transmitter handles data in the downlink direction which represents the traffic from the network to the terminal. This traffic uses several channels in the Uu interface. In the Uu interface the effective bandwidth for WCDMA is 3.84 MHz and with guard bands the required bandwidth is 5 MHz (Figure 3.19).

As mentioned earlier, in a DS-CDMA scheme, the data signal is scrambled by utilising the user-specific PN code at the transmitter side to achieve the spreading signal and at the receiver the received signal is extracted by using the same code sequence. In order to get a better understanding of the issue, we briefly discuss the *Information Theory*.

Information Theory is a mathematical model explaining the principles of signal transfer. This theory does not make any difference between different technologies involved because

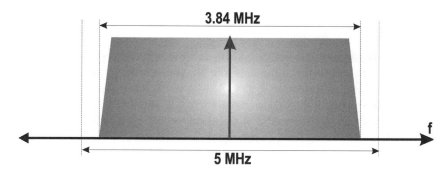

Figure 3.19. One direction of the WCDMA carrier and it's dimensions

the basic mathematics is always the same. In this context, there are several simplified principles, which can be extracted out from the Information Theory:

- The information to be transferred presents certain power, say, P_{inf}
- The wider the band for the information transfer, the smaller is the power presenting transferred information in a dedicated (small) point within the information transfer band. In other words, the total power P_{inf} is an integral over the information transfer band in this case.
- The more information there is to be transferred, the more power is required. Thus, when the power increases momentarily, P_{inf} increases, too. In this context, the higher the original bit rate to be transferred, the more power required.

If we take this into account and combine the information presented in Figure 3.11, we are able to illustrate how WCDMA treats a single originating piece of user data called a bit.

In the WCDMA air interface, every originating information bit is like a "box" having constant volume but the dimensions of the "box" change depending on the case. Referring to Figure 3.20, the depth of the "box" (frequency band) is constant in the WCDMA. The other

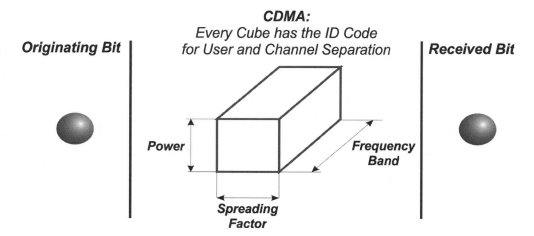

Figure 3.20. WCDMA air interface and bit treatment

two dimensions, power and spreading factor are subject to change. Based on this the following conclusions can be made:

- The better the signal can be spread, the smaller is the required energy per bit (power). This can be applied if the originating bit rate is low. In other words, spreading factor increases and power decreases.
- The smaller the spreading factor, the more energy is required per bit (power). This is applied when the originating bit rate is high. In other words, spreading factor increases and power decreases.

A confusing issue with WCDMA is that a "bit" is not a "bit" in all cases. The term bit refers to the information bit, which is a bit occurring in the original user data flow. The bit occurring in the code used for spreading is called chip. Based on this definition, we are able to present some basic items needed in WCDMA.

The bit rate of the code used for original signal spreading is, as defined, 3.84 Mb/s. This value is constant for all WCDMA variants used in 3G networks. This is called *System Chip Rate* and it is expressed as 3.84 Mcps (mega chips per second). With this system chip rate the size of one chip in time is 1/3 840 000 = 0.00000026041s.

As mentioned previously, the basic idea in WCDMA is that the signal to be transferred over the radio path is formed by multiplying the original, baseband digital signal with another signal, which has a much greater bit rate. Because both of the signals consist of bits, one must make a clear separation of the kind of bits in question (Figure 3.21). Hence:

- In the air interface the information is transmitted as *symbols*. Symbol flow is a result of modulation. Before modulation the user data flow consisting of bits has gone through channel coding, convolutional coding and rate matching. Referring to Figure 3.20, the cube in the middle of the picture actually represents 1 symbol. Depending on the used modulation method 1 symbol represents different amount of bits. In the case of DS-WCDMA-FDD, 1 symbol transmitted in an uplink direction represents 1 bit and 1 symbol

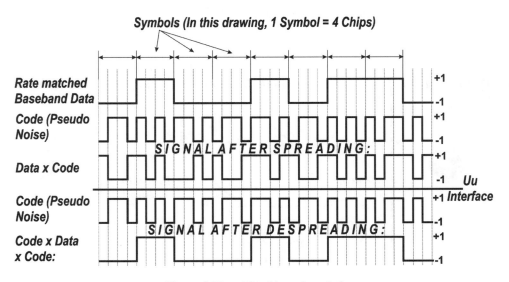

Figure 3.21.   Bit, chip and symbol

transmitted in downlink direction represents 2 bits. This difference is due to the different modulation methods used in uplink and downlink directions.
- One bit of the code signal used for signal multiplying is called a *chip*.

But how can the desired signal be captured at the receiver side? In principle, the process is quite straightforward; every receiver uses its unique code to pick up the desired signal. The received signal is multiplied by a receiver-specific code, resulting in the multiplication data. If the right code and desired received signal are multiplied, the result will be the post-integrated data with the clear peaks on the signal, otherwise the post-integrated data does not include clear signal peaks to be processed further.

The spreading factor is a multiplier describing the number many chips is used in the WCDMA radio path per 1 symbol. The spreading factor K can be expressed mathematically as follows:

$$K = 2^k$$

where $k = 0, 1, 2, ..., 8$.

For instance, if $k = 6$ the spreading factor K gets a value of 64 indicating that 1 symbol uses 64 chips in the WCDMA radio path in the uplink direction (refer to Tables 3.1 and 3.2). Another name for spreading factor is *processing gain* (G_p), and it can be expressed as a function of used bandwidths:

$$G_p = \frac{B_{Uu}}{B_{Bearer}} = \frac{\text{System chip rate}}{\text{Bearer bit rate}} = \text{Spreading factor}$$

In the formula, B_{Uu} stands for the bandwidth of the Uu Interface and B_{Bearer} is the bandwidth of the rate-matched baseband data. In other words, B_{Bearer} contains already excessive information like channel coding and error protection information. Based on the equation above and taking into account the different bit amounts 1 symbol carries in uplink and downlink directions we are able to calculate the bearer bit rates available in WCDMA.

These figures are indicative since the share of user data (payload) changes according to the used radio channel configuration.

The WCDMA system uses several codes. In theory, one type of code should be enough but in practise, the radio path physical characteristics require that the WCDMA system should use

Table 3.1 Spreading factor–symbol rate–bit rate relationship in the uplink direction

Spreading factor value	Symbol rate (ks/s)	Channel bit rate (kb/s)
256	15	15
128	30	30
64	60	60
32	120	120
16	240	240
8	480	480
4	960	960

Table 3.2   Spreading factor–symbol rate-bit rate relationship in the downlink direction

Spreading factor value	Symbol rate (ks/s)	Channel bit rate (kb/s)
512	7.5	15
256	15	30
128	30	60
64	60	120
32	120	240
16	240	480
8	480	960
4	960	1920

different codes for different purposes, and those codes should have certain features making them suitable for their use. There are basically three kinds of codes available, channelisation codes, scrambling codes and spreading code(s). Their use is shown in Table 3.3.

Table 3.3   Code types of WCDMA

	Uplink direction	Downlink direction
Scrambling codes	User separation	Cell separation
Channelisation codes	Data and control channels from the same terminal	Users within one cell
Spreading code	Channelisation code × scrambling code	Channelisation code × scrambling code

Another confusing issue with WCDMA is that the same item could have many names. This naming confusion is also present as far as codes are concerned. Scrambling code, for instance, is also known as gold code (often used in radio path related technical publications) and long code. From these alternatives, the name scrambling code is the preferred name. The scrambling code is used in the downlink direction for cell/sector separation. Scrambling codes are used also in the uplink direction. In this case, the users (mobiles) are separated from each other with this code.

In the case of downlink, a total of $2^{18} - 1 = 262\,143$ scrambling codes, numbered $0\ldots262\,142$ can be generated. However not all the scrambling codes are used. The scrambling codes are divided into 512 sets each having a primary scrambling code and 15 secondary scrambling codes (Figure 3.22).

The primary scrambling codes consist of scrambling codes $n = 16*i$ where $i = 0,\ldots,511$. The ith set of secondary scrambling codes consists of scrambling codes $16*i + k$, where $k = 1,\ldots,15$. There is a one-to-one mapping between each primary scrambling code and 15 secondary scrambling codes in a such a way that the ith primary scrambling code corresponds to ith set of scrambling codes.

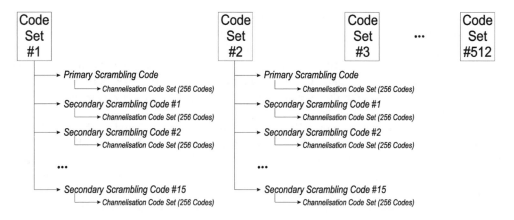

- 512 Code Sets x 16 Scrambling Codes = 8192 Codes numbered from 0 ... 8191 available

Figure 3.22.   Scrambling and channelisation codes; amounts and relationship

Hence, according to the above, there are $k = 0, 1,...,8191$ scrambling codes available. Based on 3GPP TS 25.213,each of these codes is associated with an even alternative scrambling code and an odd alternative scrambling code that may be used for compressed frames. The even alternative scrambling code corresponding to scrambling code k is scrambling code number $k + 8192$, while the odd alternative scrambling code corresponding to scrambling code k is scrambling code number $k + 16384$. The set of primary scrambling codes is further divided into 64 scrambling code groups, each consisting of eight primary scrambling codes. The jth scrambling code group consists of primary scrambling codes $16*8*j + 16*k$, where $j = 0,...,63$ and $k = 0,...,7$.

Each cell is allocated one and only one primary scrambling code. The primary CCPCH (Common Control Physical Channel) is always transmitted using the primary scrambling code. Hence, the other downlink physical channels can be transmitted with either the primary scrambling code or a secondary scrambling code from the set associated with the primary scrambling code of the cell.

For the uplink case the situation is different because there are millions of uplink scrambling codes available. The specified amount of uplink scrambling codes for WCDMA is 2^{24}. All uplink channels shall use either short or long scrambling codes, except the PRACH (Physical Random Access Channel), for which only the long scrambling code is used. Therefore, uplink code allocation is not as crucial as downlink code allocation in WCDMA.

Channelisation codes are used for channel separation both in uplink and downlink directions. Channelisation codes are the ones that have different spreading factor values and thus also different symbol rates. There is a total of 256 pieces of channelisation codes available and the spreading factor indicates how many bits of those codes are used in the connection. Thus, the greater the spreading factor value is, the better the channelisation codes are utilised and the radio resources are used in an optimal way. In the case of high user data rates the spreading factor gets a relatively small value. Respectively, this leads to the situation where high user data rates consume more air interface code capacity.

As is their nature, the channelisation codes are *orthogonal*, or, one can say, they have orthogonal properties. Orthogonality as a term means that the channelisation codes in the 256-member code list are selected in a way that they interfere with each other as little as possible. This is necessary in order to have a good channel separation. On the other hand, the code used for user and cell separation (scrambling code) must have good correlation properties. The channelisation codes do have those and this is the basic reason why both scrambling and channelisation codes are used.

Every WCDMA cell uses normally one downlink scrambling code, which is locally unique and acts basically like a cell ID. The characteristics of this scrambling code are pseudo-random, i.e. it is not always orthogonal. Under this scrambling code the cell has a set of channelisation codes, which is orthogonal in nature and used for channel separation purposes.

In WCDMA, the users transmitting and receiving use the whole available frequency band simultaneously in time. To separate different transmissions spread over the frequency band, spreading codes are used. A spreading code is a unique code assigned to the beginning of the transaction by the network. A spreading code can be imagined to be like a "key" which is used both by the mobile and the network. Both ends of the connection use this "key" to open the noise-like transmitted wideband signal. Or to be exact, to extract the correct wideband transmission away from the frequency band, since the transmitted wideband signal may contain many mobile-network connections.

From the point of view of the spreading code, the capacity of a cell depends on the downlink scrambling code amount assigned for the cell (minimum is 1). Every downlink scrambling code then has a set of channelisation codes under it and every call/transaction requires one channelisation code to operate. In practise, one spreading code is actually *scrambling code × channelisation code*. If channelisation codes are not used, the spreading code is the same as the scrambling code. Also, the spreading code depends on the information type to be delivered. The information being common in nature does not use channelisation code. For example, the broadcast information the cell delivers downlink is one of those.

3.5.3. WCDMA Radio Channels

The WCDMA radio access allocates bandwidth for users and the allocated bandwidth and its controlling functions are handled with the term "Channel". The functionality implemented through the WCDMA defines what kinds of channels are required and how they are organised. The channel organisation the WCDMA uses is a three-layer one; there are logical channels, transport channels and physical channels. From these, the logical channels describe the types of information to be transmitted, transport channels describe how the logical channels are to be transferred and the physical channels are the "transmission media" providing the radio platform through which the information is actually transferred (Figure 3.23).

Referring to the architecture issues explained in Chapter 1, the channel structures and their use differs remarkably from the GSM. The term physical channels means different kinds of bandwidths allocated for different purposes over the Uu interface. In other words, the physical channels actually form the physical existence of the Uu interface between the UE domain and access domain. In GSM the physical channels and their structure is recognised by the BSC but in WCDMA the physical channels really exist in the Uu interface and the RNC is not necessarily aware of their structure at all.

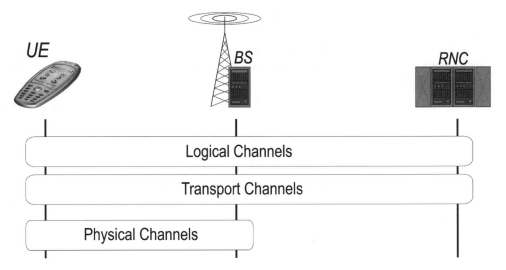

Figure 3.23.   Channel types and their location in UTRAN

   Instead of physical channels the RNC "sees" transport channels. Transport channels carry different information flows over the Uu interface and the physical element mapping these information flows to the physical channels is the BS (Base Station). Logical channels are not actually channels as such, rather they can be understood as different tasks the network and the terminal should perform in different moments of time. These partially timely structures are then mapped into transport channels performing actual information transfer between the UE domain and access domain.

   Concerning the logical channels, the UE and the network have different tasks to do. Thus, the logical, transport and physical channel structures are a bit different in either direction. Roughly, the network has the following tasks to perform:

- The network must inform the UE about the radio environment. This information consists of, for instance, the code value(s) used in the cell and in the neighbouring cells, allowed power levels, etc. This kind of information the network provides for the UE through the logical channel called the Broadcast Control Channel (BCCH).
- When there is a need to reach a UE for communication (for instance, a mobile terminated call) the UE must be paged in order to find out its exact location. This network request is delivered on the logical channel called the Paging Control Channel (PCCH).
- The network may have certain tasks to do, which are or may be common for all the UE residing in the cell. For this purpose the network uses the logical channel called Common Control Channel (CCCH). Since there could be numerous UE using the CCCH simultaneously, the UE must use U-RNTI (UTRAN Radio Network Temporary Identity) for identification purposes. By investigating the received U-RNTI the UTRAN is able to route the received messages to the correct serving RNC. U-RNTI is discussed in Chapter 4.
- When there is a dedicated, active connection, the network sends control information concerning this connection through the logical channel called Dedicated Control Channel (DCCH).

- Dedicated traffic: the dedicated user traffic for one user service in the downlink direction is sent through the logical channel called Dedicated Traffic Channel (DTCH).
- Common Traffic Channel (CTCH) is an unidirectional channel existing only in the downlink direction and it is used when transmitting information either to all UE or a specific group of UE in the cell.

The transport channels shown in Figure 3.24 except one are mandatory ones. The mandatory transport channels are Broadcast Channel (BCH), Paging Channel (PCH), Forward Access Channel (FACH) and Dedicated Channel (DCH). In addition to these, the operator may configure the UTRA to use the Downlink Shared Channel (DSCH). In these transport channels, the only dedicated transport channel is DCH; the others are common ones. In this context the term "dedicated" means that the UTRAN has allocated the channel to be used between itself and certain terminals. The term "common" means that several terminals could use the channel simultaneously.

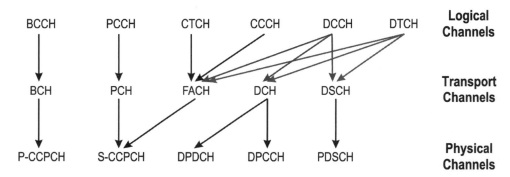

Figure 3.24. Logical, transport and physical channel mapping in the downlink direction

The BCH carries the content of BCCH, i.e. the UTRA specific information to be delivered in the cell. This information consists of, for example, random access codes, access slot information and information about the neighbouring cells. The UE must be able to decode the BCH in order to register to the network. The BCH is transmitted with relatively high power because every terminal in the intended cell coverage area must be able to "hear" it. The PCH carries paging information. The PCH is used when the network wishes to initiate a connection with certain UE. FACH carries control information to the UE known to be in the cell. For example, when the RNC receives a random access message from the terminal, the response is delivered through FACH. In addition to this, the FACH may carry packet traffic in the downlink direction. One cell may contain numerous FACH but one of these is always configured in such a way (with low bit rate) that all the terminals residing in the cell area are able to receive it. The DCH carries dedicated traffic and control information, i.e. the DCCH and DTCH. It should be noted that one DCH might carry several DTCH depending on the case. For example, a user may have a simultaneous voice call and video call active. The voice call uses one logical DTCH and the video call requires another logical DTCH. Both of these, however, use the same DCH. From the UTRA capacity point of view, the aim is to use common transport channels as much as possible, since the dedicated channels will occupy radio resources. The optional

DSCH is a target of increasing interest. It carries dedicated user information (DTCH and DCCH) for packet traffic and several users can share it. In this respect it is better than DCH, since it saves packet traffic related network resources in the downlink direction. Another point is that the maximum bit rate for DSCH can be changed faster than in DCH. The expected wide use of services producing occasional packet bursts, like for instance web surfing, have increased the interest towards DSCH.

In the uplink direction the logical channel amount required is smaller. There are only three logical channels, CCCH, DTCH and DCCH. These abbreviations have the same meaning as in the downlink direction (Figure 3.25).

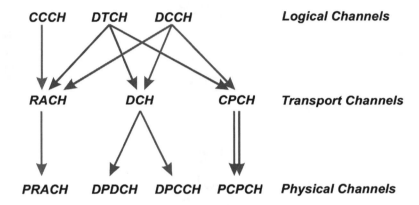

Figure 3.25.   Logical, transport and physical channel mapping in the uplink direction

There are three mandatory transport channels in the uplink direction being Random Access Channel (RACH), DCH and Common Packet Channel (CPCH). The RACH carries control information from the UE to the UTRAN, like for instance, connection set-up requests. In addition to this the RACH may carry small amounts of packet data. The DCH is the same as in the downlink direction, i.e. dedicated transport channel carrying DCCH and DTCH information. The CPCH is a common transport channel meant for packet data transmission. It is a kind of extension for RACH and the CPCH counterpart in the downlink direction is FACH.

When the information is collected from the logical channels and organised to the transport channels it is in ready-to-transfer format. Before transmitting the transport channels are arranged to the physical channels. The Uu interface contains more physical channels than Figs. 24 and 25 indicate. The other physical channels present are used for physical radio media control, modification and access purposes. The overview of WCDMA physical channels is provided in Figure 3.26.

The physical channels are used between the terminal and the base station. Due to the network architecture solution explained at the beginning of this book, the physical access (i.e. the physical channels) is separated from the other layers. This arrangement makes it possible to swap the physical radio access medium below the other layers, in theory. In practise the radio access medium change will reflect on higher layers but this arrangement minimises those changes.

The Primary Common Control Physical Channel (P-CCPCH) carries the BCH in the downlink direction. The P-CCPCH is available in a way that all the terminals populated

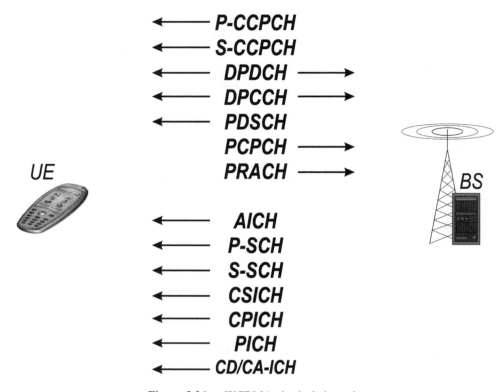

Figure 3.26.   WCDMA physical channels

within the cell coverage are able to demodulate its contents. Because of this requirement, the P-CCPCH has some characteristics, which actually are limitations when compared to the other physical channels in the system. The P-CCPCH uses a fixed channelisation code and thus its spreading code is fixed, too. This is a must because otherwise the terminals are not able to "see" and demodulate the P-CCPCH. The P-CCPCH bit rate is 30 kb/s with a spreading factor value of 256. The bit rate must be low because this channel is transmitted with relatively high power. If using higher bit rates, the interference starts to increase thus limiting the system capacity.

The Secondary Common Control Physical Channel (S-CCPCH) carries two transport channels in it: Paging Channel (PCH) and Forward Access Channel (FACH). These transport channels may use the same or separate S-CCPCH, thus a cell always contains at least one S-CCPCH. The bit rate of S-CCPCH is fixed and relatively low due to the same reasons concerning the P-CCPCH. At a later phase the S-CCPCH bit rate can be increased by changing system definitions. The configuration of S-CCPCH is variable: depending on the case, the S-CCPCH can be configured differently in order to optimise the system performance. For instance, the pilot symbols can be included or not. Referring to the variable configuration alternatives of the S-CCPCH, one alternative to increase system performance is to multiplex the PCH information together with the FACH to the S-CCPCH and the PCH related paging indications to a separate physical channel called a Paging Indicator Channel (PICH).

The Dedicated Physical Data Channel (DPDCH) carries dedicated user traffic. The size of the DPDCH is variable and it may carry several calls/connections in it. As the name says, it is a dedicated channel, which means that it is used between the network and *one* user. The dedicated physical channels are always allocated as pairs for one connection: one channel for control information transfer and the other for actual traffic. The DPCCH transfers the control information during the dedicated connection. Figure 3.27 shows how the DPDCH and the DPCCH are handled in both uplink and downlink directions.

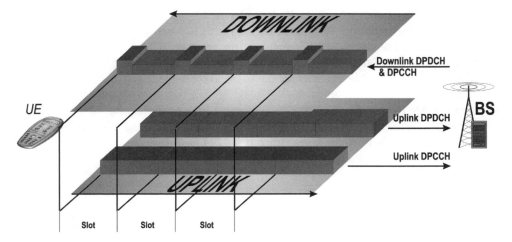

Figure 3.27.   DPDCH and DPCCH in uplink and downlink directions

In the downlink direction the DPDCH (carrying user data) and DPCCH (carrying for example power control data rate information) are time multiplexed. If there is nothing to carry in DPDCH the transmitted signal is pulse-like causing EMC type disturbances, this is not a problem in the downlink direction. In the uplink direction the DPDCH and DPCCH are separated by I/Q modulation. If there is no user data to carry in DPDCH, no pulse-like disturbances exist either. The outcome of the I/Q modulation in the UE is actually one channel but carrying two information branches (see Figure 3.27) and both of the information branches consume one code resource.

Together, the DPCCH and the DPDCH carry the contents of the transport channel DCH. When the dedicated connection uses a high peak bit rate the system easily starts to suffer from the lack of channelisation codes in the cell. In this case, there are two options: either to add a new scrambling code to the cell or to use common channels for the dedicated data transmission. The adding of the scrambling codes is not recommended because the orthogonality is lost. Instead, using common channel resources for packet data transmission is seen a better way to increase capacity. The downlink DCH is able to provide information whether the receiving UE must decode the Physical Downlink Shared Channel (PDSCH) for additional user information. The PDSCH carries the DSCH transport channel and as it was explained earlier, the DSCH is an optional feature the operator may configure to be used or not.

If there is a need to send packet data in an uplink direction and the RACH packet transfer capacity is not enough, the UE may use CPCH (uplink Common Packet Channel). The correspondent physical channel in the uplink direction is PCPCH (Physical (uplink) Common Packet Channel). The counterpart of the CPCH in the downlink direction is DPCCH.

The PRACH carries information related to the Random Access Procedure (Figure 3.28). With this procedure the terminal accesses to the network and also small data amounts can be transferred. The Random Access Procedure has the following phases:

1. The UE decodes the BCH information on the P-CCPCH and finds out the RACH slots available as well as the scrambling code(s).
2. The UE selects randomly one RACH slot to be used.
3. The UE sets the initial power level to be used (this is based on the received downlink power level) and sends so-called preamble to the network.
4. The terminal decodes the AICH (Acquisition Indication Channel) to find out if the network notified the sent preamble. If it was not the UE sends preamble again but with a higher bit power level.
5. When the AICH indicates that the network notified the preamble, the UE sends RACH information on the PRACH. The length of the RACH information sent is either one or two WCDMA frames, in time this takes either 10 or 20 ms.

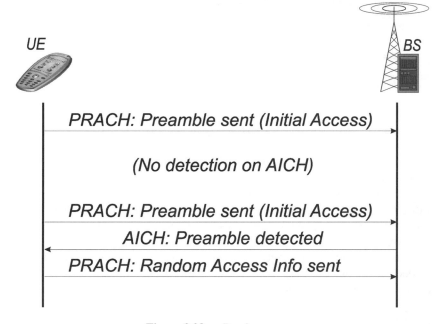

Figure 3.28.   Random access

The synchronisation channel provides the cell search information for the UE residing the cell coverage area. The SCH is actually a combination of two channels, Primary Synchronisation Channel (P-SCH) and Secondary Synchronisation Channel (S-SCH). The P-SCH uses a fixed channelisation code, whose length is 256 and this channelisation code is the same in every cell of the system. When the UE has demodulated the P-SCH it has achieved frame and slot synchronisation of the system and it also knows to which scrambling code group the cell to be accessed belongs.

The Common Pilot Channel (CPICH) is an unmodulated code channel which is scrambled with the cell specific scrambling code. The CPICH is used for dedicated channel estimation (by the terminal) and to provide channel estimation reference(s) when common channels are concerned. In this respect, the pilot signal is somewhat the same with the same functionalities as the training sequence included in the middle of a GSM burst. Normally, one cell has only one CPICH but there can be two of them. In this case, those channels are called primary CPICH and secondary CPICH. The cell may contain the secondary CPICH, for instance, when the cell contains narrow beam antenna aiming to offer service for a dedicated "hot spot" area. Thus, a dedicated area uses the secondary CPICH and the primary CPICH offers pilot for the whole cell coverage area. The terminals listen to the pilot signal continuously and this is why it is used for some "vital" purposes in the system, e.g. handover measurements and cell load balancing: From the system point of view, the CPICH power level adjustment balances load between cells: the UE always searches the most attractive cells and by decreasing the CPICH power level the cell is less attractive.

The other physical channels listed in Figure 3.26 are CPCH Status Indication Channel (CSICH), CPCH Collision Detection Indicator Channel (CD-ICH) and CPCH Channel Assignment Indicator Channel (CA-ICH). The CSICH uses free space occurring in AICH and the CSICH is used to inform the UE about CPCH existence and configuration. To avoid collisions (two UE using same identity patterns), a CD-ICH and CA-ICH are used. These physical channels transfer the collision detection information towards the UE.

3.5.4. WCDMA Frame Structure

In order for the radio access to handle the control actions like timing, synchronisation arrangements, transmission assurance, etc. between the network and mobile station the burst data should be structured in a well-defined way. Therefore, the WCDMA contains a frame structure, which is divided into 15 slots, each of length 2/3 ms and thus the frame length is 10 ms (Figure 3.29).

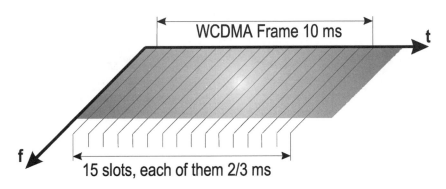

Figure 3.29. WCDMA frame

Based on this, one WCDMA frame is able to handle

$$\frac{0.010 \text{ s}}{0.00000026041 \text{ s}} \approx 38400 \text{ Chips}$$

One slot in the WCDMA frame contains

$$\frac{38400 \text{ Chips}}{15 \text{ Slots}} = 2560 \text{ Chips}$$

Unlike in GSM, WCDMA does not contain any super-, hyper- or multiframe structures. Instead, WCDMA frames are numbered by an SFN (System Frame Number). An SFN is used for internal synchronisation of UTRAN and timing of BCCH information transmission.

Figures 3.30 and 3.31 illustrate the uplink and downlink dedicated channel frame structures, respectively. As shown in the figures, the dedicated physical channels have different structure for uplink and downlink.

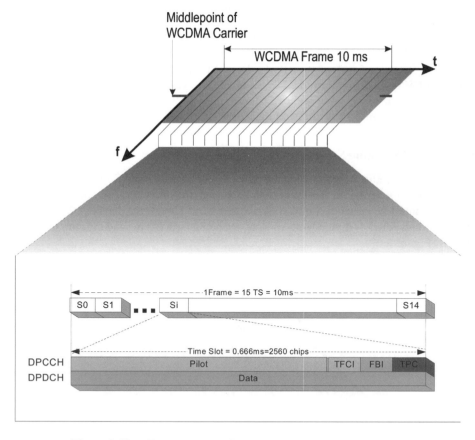

Figure 3.30.   Frame structure for a UL dedicated physical channel

In the uplink case, the basic frame structure of the dedicated physical channel follows the downlink frame structure, but the main difference is that uplink dedicated channels cannot be

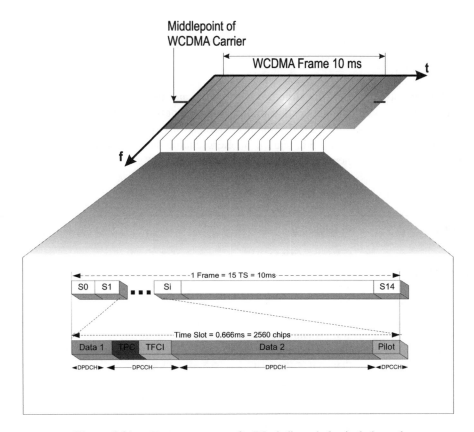

Figure 3.31. Frame structure of a DL dedicated physical channel

considered as a time multiplex of DPDCH and DPCCH. This leads to the fact that multicode operation is possible for the UL dedicated physical channels. It should be mentioned, however, when multicode is used, several parallel DPDCHs are transmitted using different channelisation codes with only one DPCCH per connection.

The uplink DPCCH consists of pilot bits to support channel estimation for coherent detection, Transmit Power Control (TPC) commands for transmit power adjusting, Feedback Information (FBI), and an optional Transport Format Combination Indicator (TFCI) to inform the receiver about the instantaneous parameters of the different transport channels multiplexed on the uplink DPDCH, and corresponds to the data transmitted in the same frame.

In the case of DL, every slot includes pilot bits, transmitted power control bits, a transport frame indicator, and data.

It needs to be mentioned that within one downlink DPCH, the dedicated data generated at layer 2 and above, i.e. the DCH, is transmitted in time multiplex with control information generated by layer 1 (known pilot bits, TPC commands, and an optional TFCI). Therefore, the downlink DPCH can be seen as a timed multiplex of downlink DPDCH and downlink DPCCH. In addition to the harmonisation compromise in standardisation, the reasons behind this approach are:

- To minimise the continuous transmission by mobile handset;
- To use the downlink orthogonal codes more efficiently;
- To minimise the power control delay by utilising the slot-offset between uplink and downlink slots.

For the common channels the structure is the same and the main difference between the common and dedicated channels is that in common channels the TPC bits are not used.

Finally, Table 3.4 summarises the main characteristics of the WCDMA-FDD scheme described previously. These technical parameters enable the WCDMA-FDD scheme to meet the system requirements for the 3G mobile system.

Table 3.4   WCDMA-FDD: main technical characteristics

Parameter	Specification
Multiple access	FDD: DS-CDMA
Duplex scheme	FDD
Chip rate (Mc/s)	3.84
Frame length (ms)	10
Channel coding	Convolutional coding ($R = 1/2$, $1/3$, $1/4$, $K = 9$); turbo code of $R = 1/2$, $1/3$, $1/4$ and $k = 4$
Interleaving	Inter/intraframe
Data modulation	FDD: DL:QPSK, UL dual channel QPSK
Spreading modulation	FDD: UL:BPSK, DL:QPSK
Power control	Closed loop (inner loop, and outer loop), open loop. Step size: 1–3 dB (UL); power cycle: 1500/s
Diversity	RAKE in both BS and MS; antenna diversity; transmit diversity
Inter-BS synchronisation	FDD: no accurate synchronisation needed
Detection	MS&BS: pilot symbol based coherent detection in UL, CPICH cannel estimation in DL
Multiuser detection	Supported (not at the first phases)
Service multiplexing	Variable mixed services per connection is supported
Multirate concept	Is supported by utilising a variable spreading factor and multicode
Handover	Intra-frequency soft and softer handovers are supported, inter-system and inter-frequency handovers are supported

4

UMTS Radio Access Network (UTRAN)

The main task of UTRAN is to create and maintain Radio Access Bearers (RAB) for communication between User Equipment (UE) and the Core Network (CN). With RAB the CN elements are given an illusion about a fixed communication path to the UE, thus releasing them from the need to take care of radio communication aspects. Referring to the network architecture models presented in Chapter 1, UTRAN realises certain parts of Quality of Service (QoS) architecture independently (Figure 4.1).

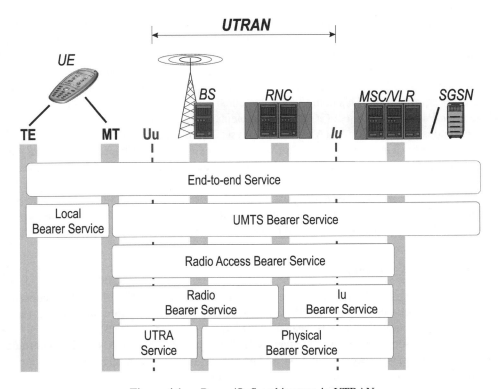

Figure 4.1. Bearer/QoS architecture in UTRAN

UTRAN is located between two open interfaces being Uu and Iu. From the bearer architecture point of view, the main task of UTRAN is to provide bearer service over these interfaces; in this respect the UTRAN controls Uu interface and in Iu interface the bearer service provision is done in co-operation with the CN.

The RAB fulfils the QoS requirements set by the CN. The handling of end-to-end QoS requirement in the CN and in the UE is the responsibility of Communication Management. Those requirements are then mapped onto the RAB, which is "visible" for the (Mobile Termination) MT and the CN. As said earlier, the main task of UTRAN is to create and maintain RAB so that the end-to-end QoS requirements are fulfilled in all respects.

One of the main ideas of this layered structure is to encapsulate the physical radio access, later it can be modified or replaced without changing the whole system. In addition, it is a known fact that the radio path is very complex and continually changing transmission media. This bearer architecture gives remarkable role to the Radio Network Controller (RNC), since the RNC and the CN map the end-to-end QoS requirements over the Iu interface and the RNC takes care of satisfying the QoS requirements over the radio path. These two bearers exist in the system because the Iu bearer is more stable in nature; the RB experiences more changes during the connection. For example, one UE may have three continuously changing RBs maintained between itself and the RNC, still the RNC has only one Iu bearer for this connection. This kind of situation occurs in context with soft handover, which is described in this chapter.

The physical basis of the end-to-end service in Uu interface is Universal Terrestrial Radio Access (UTRA) Service. As was explained in Chapter 3, UTRA Service is implemented with Wideband Code Division Multiple Access (WCDMA) radio technology and in the beginning of UMTS the used variant was WCDMA-FDD. From the QoS point of view the UTRA Service contains mechanisms that show how the end-to-end QoS requirements are mapped onto the physical radio path. Respectively, every Uu interface connection requires terrestrial counterpart through UTRAN. The equal terrestrial physical basis for the end-to-end service is Physical Bearer Service. This could be implemented with various technologies but in Third Generation Partnership Project (3GPP) R99 implementation the most probable alternative is Asynchronous Transfer Mode (ATM) over physical transmission media. Both ATM and some of the physical transmission alternatives are explained in Chapter 9. As UTRAN evolves the ATM will have Internet Protocol (IP) as an alternative way of implementation (refer to Chapter 2)

Referring to the conceptual model presented in Chapter 1, the protocols realising the RB Service belong to Access Stratum. The protocols above RB Service belong to Non-Access Stratum and are responsible for the UMTS bearer service.

4.1. UTRAN Architecture

Figure 4.2 illustrates the UTRAN architecture on the network element level. UTRAN consists of Radio Network Subsystems (RNS) and each RNS contains various amount of Base Stations (BS, or officially Node B) realising the Uu interface and one RNC.

The RNSs are separated from each other by UMTS Interface between RNCs (Iur) interface forming connections between two RNCs. The Iur, which has been specified as an open interface, carries both signalling and traffic information.

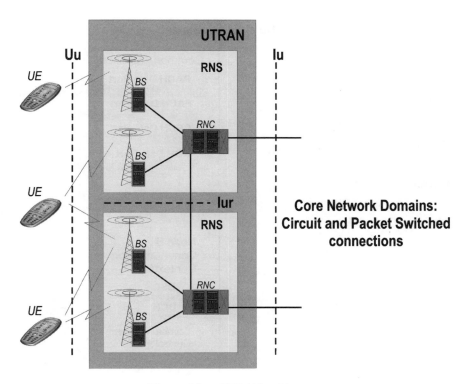

Figure 4.2. UTRAN architecture

4.2 Base Station

The Base Station (BS) is located between the Uu and UMTS Interface between RNC and BS (Iub) interfaces. Its main tasks are to establish the physical implementation of the Uu interface and, towards the network, the implementation of the Iub interface by utilising the protocol stacks specified for these interfaces. Realisation of the Uu interface means that the BS implements WCDMA radio access physical channels and transfers information from transport channels to the physical channels based on the arrangement determined by the RNC (for channels and their descriptions, refer to Chapter 3).

4.2.1 Base Station Structure

The internal structure of the BS is a vendor-dependent issue, but the logical structure, i.e. how a BS is treated within UTRAN, is generic. From the network point of view, the BS can be divided into several logical entities as shown in Figure 4.3.

On the Iub side, a BS is a collection of two entities, common transport and a number of Traffic Termination Points (TTP). The common transport represents those transport channels that are common for all UE in the cell and those used for initial access. The common transport entity also contains one Node B Control port used for operation and maintenance (O&M) purposes. One TTP consists of a number of *Node B Communication Contexts*. Node B Communication Context in turn consist of all dedicated resources required when the UE is

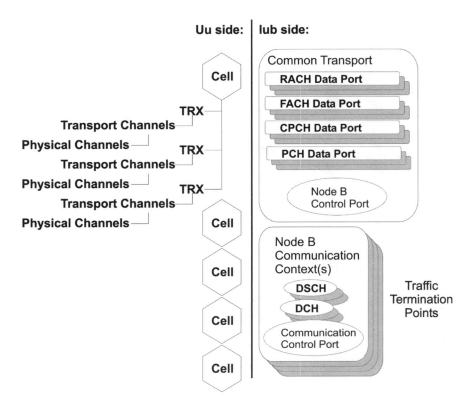

Figure 4.3.   BS logical structure

in dedicated mode. Thus, one Node B Communication Context may contain at least one Dedicated Channel (DCH), for example. The exception is the Downlink Shared Channel (DSCH), which also belongs to Node B Communication Context. From the point of view of UMTS network infrastructures, the BS can be considered to be a logical O&M entity, which is subject to network management operations. In other words, this term describes physical BS and its circumstances, BS Site.

From the point of view of the radio network and its control, BS consists of several other logical entities called cells. A cell is the smallest radio network entity having its own identification number (cell ID), which is publicly visible for the UE. When the radio network is configured, it is actually the cell(s) data that is changed. The term sector stands for the physical occurrence of the cell, i.e. radio coverage.

Every cell has one scrambling code, UE recognises a cell by two values, scrambling code (when logging into a cell) and cell ID (for radio network topology). One cell may have several Transmitter-Receivers (TRXs, also called carriers) under it. The TRX of the cell delivers the broadcast information towards the UE. This is, the Primary Common Control Physical Channel (P-CCPCH) containing the Broadcast Channel (BCH) information is transmitted here. One TRX maintains physical channels through the Uu interface and these carry the transport channels containing actual information, which may be either common or dedicated in nature.

A cell may consist a minimum of one TRX. The TRX is physically a part of the BS performing various functions, data flows are converted from terrestrial Iub connection to the radio path and vice versa.

4.2.2 Modulation Method

From the system point of view, the used modulation method is of interest since it has a close relationship to the overall system capacity and performance. WCDMA uses Quadrature Phase Shift Keying (QPSK) as its modulation method in the downlink direction. To make it simple, this modulation method expresses a bit and its status with a different phase of the carrier. The bits in the modulation process are handled as pairs and thus, this generates four possible two-bit-combinations to be indicated.

In the modulation process, the data stream to be modulated, i.e. physical channels are first converted from serial to parallel format. After this conversion, the modulation process divides the data stream into two branches, I and Q. In the I branch the bit having value "1" represents + 180° phase shift and value "0" means the phase of the carrier has not shifted. In the Q branch the bit having value "1" represents + 90° phase shift and value "0" stands for − 90° phase shift. These branches, I and Q, are fed by the oscillator, I branch directly and Q branch with 90° shift. When the incoming data stream is combined with the oscillator output, the two-bit-combinations (one bit from I and one from Q branch) represent the phase shifting shown on the right-hand side of Figure 4.4. Thin lines in the square indicate the "paths" where the system can transfer itself from one state (phase) to another.

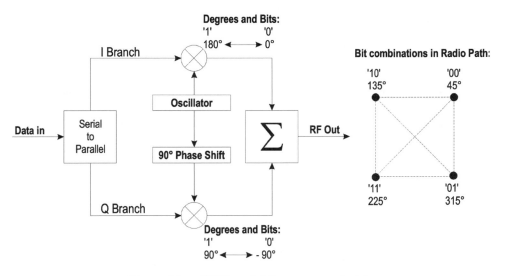

Figure 4.4.   QPSK modulation process principle

This system works fine, but certain two-bit-combinations are more difficult to handle. For instance, when the bit values are changing from "00" to "11" it means 180° phase shifting which can be seen as very remarkable amplitude change. Very big amplitude changes cause problems especially if the bandwidth used in the radio connection is wide. In this case the BS must have linear amplifiers in order to guarantee that the amplitude changes are represented correctly throughout the used bandwidth.

To eliminate this too-rapidly-changing-amplitude problem, another variant of the QPSK is used: Offset Quadrature Phase Shift Keying (OQPSK). OQPSK introduces a time delay on the Q branch of the modulator and this time delay is equal to 0.5 bits (chips). This prevents 180° phase changes and limits them to 90° steps. Based on this, the transition from the combination "11" to "00" is now "11" − "10" − "00" and within the same time as in the QPSK (Figure 4.5).

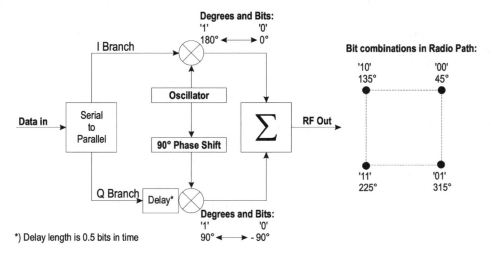

Figure 4.5.   OQPSK modulation principle

The result is that the spectrum used for QPSK and OQPSK are the same but OQPSK has smoother signal. This allows the amplifiers to operate also on their non-linear operating area without problems. Based on these facts, the WCDMA actually uses both QPSK variants, the conventional QPSK in the downlink direction and the OQPSK in the uplink direction (Figure 4.6).

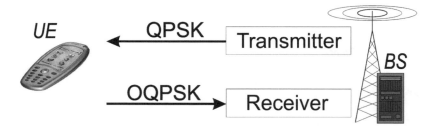

Figure 4.6.   Used modulation methods and their directions

The conventional QPSK could be used in both directions but then the UE would suffer power consumption problems and high prices: the linear power amplifier should be very accurate and thus very expensive. With OQPSK these amplifying problems could be avoided. From the BS point of view, the OQPSK is not "good enough" because the signal the BS transmits must be very accurate and the UE must achieve accurate synchronisation. With OQPSK, these features are compromised and thus the conventional QPSK is better for modulation performed on the transmission side of the BS.

4.2.3 Receiver Technique

The WCDMA utilises multipath propagation. This is, the transmitted signal propagates in many different ways from the transmitter to the receiver. On the other hand, to gain better capacity in the radio network, the transmit powers of the UE (and BSs) should be relatively small. This decreases interference in the radio interface and gives more space for other transmissions. This leads to a situation where it is very useful that both the UE and the BS are able to "collect" many weak level signals indicating the same transmission and to combine them together. This requires a special type of receiver. One example of this kind of arrangement is called RAKE. It should be noted that RAKE is not an acronym; it is real name for this type of receiver.

Therefore, the purpose of the RAKE receiver is to improve the received signal level by exploiting the multipath propagation characteristics of the radio wave, as signals propagating by different paths have different attenuation. A basic RAKE receiver consists of a number of fingers, a combiner, a matched filter, and a delay equaliser. Practically every finger can receive part of the transmitted signal, which can be either from the serving BS or from neighbouring BS. Different branches of the received signal are combined in such a way that the phase and amplitude deviations of the branches compensate for each other, resulting in a signal with remarkably higher signal strength than every individual signal branch. In order for the multipath signal branches to be distinguished from each other, there must be a specific delay range between consecutive branches.

In general, diversity techniques are efficient means to overcome the radio signal deterioration due to shadowing and fading which is discussed in Chapter 3. In addition, utilising diversity technique is a prerequisite for providing soft handover feature in cellular systems. Different diversity techniques may be used in mobile systems, that is, time diversity, space diversity, frequency diversity, etc. In WCDMA technology, typically polarisation diversity is utilised for both the uplink and downlink transmission. The purpose of multipath diversity is to resolve individual multipath components and combine them to obtain a sum signal component with better quality. Having used the RAKE receiver both in UE and the BS makes it possible to capture and combine different branches of the desired signal and improve the quality of the ultimate signal or data stream to be processed further. Maximum Ratio Combining (MRC) is the diversity algorithm used in the signal processing. The responsibility for diversity is distributed in radio access between UE, BS, and RNC depending on the link direction (uplink, downlink) and the position of the network element in the system architecture hierarchy.

4.2.4 Cell Capacity

In WCDMA technology, all the users share the common physical resource, frequency band in 5 MHz slices. All users of the WCDMA TRX co-exist on the frequency band at the same moment of time and different transactions are recognised with spreading codes. In the first UMTS systems, the UTRAN will use the WCDMA-FDD variant. In this variant, the Uu interface transmission directions use separate frequency bands. One of the most interesting questions (and one of the most confusing items) for people is the capacity of the WCDMA TRX. In Global System for Mobile communications (GSM) the TRX capacity calculation is a very straightforward procedure, but because in WCDMA the radio interface is handled differently and the system capacity is limited by variable factors, the capacity of the WCDMA TRX is not very easy to be determined.

The following text presents an idea of how to *roughly* estimate WCDMA TRX capacity theoretically and based on radio conditions. In order to simplify the issue, one must make some assumptions:

- All the subscribers under the TRX coverage area are equally distributed so that they have equal distances to the TRX antenna.
- The power level they use is the same and thus the interference they cause is on the same level.
- Subscribers under the TRX use the same base-band bit rate, i.e. also the same symbol rates.

Under these circumstances a value called Processing Gain (G_p) can be defined. Processing Gain is a relative indicator informing what is the relationship between the whole bandwidth available (B_{RF}) and the base-band bit rate ($B_{Information}$).

$$G_p = \frac{B_{RF}}{B_{Information}}$$

There is another way to express the G_p by using chip and data rates:

$$G_p = \frac{Chiprate}{Datarate}$$

Both ways (when announced in dB values) give as a result the improvement of the Signal to Noise Ratio (S/N) between the received signal and the output of the receiver.

Further on, the processing gain is actually the same as the spreading factor. It should be noted that, the base-band bit rate discussed here is the one achieved after the rate matching. In this process the original (user) bit rate is adjusted to the bearer bit rate. Bearer bit rates are somewhat fixed, e.g. 30 kb/s, 60 kb/s, 120 kb/s, 240 kb/s, 480 kb/s and 960 kb/s. The system chip rate is constant; 3.84 Mcps (3 840 000 chips/s). Hence, as an example the bearer having a bit rate 30 kb/s will have a spreading factor 128:

$$G_p = \frac{3\,840\,000}{30\,000} = 128 = Spreading\ Factor$$

The power P required for information transfer in one channel is a multiple of the energy used per bit and the base-band data rate.

$$P = E_b \times Baseband\ Datarate$$

On the other hand, it is known that the noise on one channel ($N_{Channel}$) using partially the whole bandwidth B_{RF} can be expressed as:

$$N_{\text{Channel}} = B_{\text{RF}} \times N_{\text{O}}$$

where N_{O} is the Noise Spectral Density (W/Hz)

Based on this, the signal to noise ratio is:

$$S/N = \frac{P}{N_{\text{Channel}}} = \frac{E_b \times \text{Baseband Datarate}}{B_{\text{RF}} \times N_{\text{O}}} = \frac{E_b/N_{\text{O}}}{G_p}$$

If assumed that there are X users under the TRX and the assumptions presented earlier in this chapter are applied, it means that there are (statistically thinking) $X - 1$ users causing interference to one user. This also indicates signal to noise ratio and when expressed in mathematical format the outcome is the following equation:

$$S/N = \frac{P}{P \times (X - 1)} = \frac{1}{X - 1}$$

If, further on, there are plenty of users (tens of them) then the equation could be simplified:

$$S/N = \frac{1}{X - 1} \approx \frac{1}{X}$$

Now there are two different ways to calculate signal to noise ratio:

$$\frac{E_b/N_{\text{O}}}{G_p} \approx \frac{1}{X} \Rightarrow X \approx \frac{G_p}{E_b/N_{\text{O}}}$$

This equation is a *very rough* expression and should be used for estimation purposes only. The "official" way to calculate TRX capacity has several more parameters to be taken into account.

Example

Assume that the spreading factor used in the cell is 128 and for those transactions the E_b/N_{O} is 3 dB. How many users can the cell contain simultaneously assuming that there is 1 TRX available?

$$X \approx \frac{G_p}{E_b/N_{\text{O}}} = \frac{128}{3 \text{ dB}} \approx \frac{128}{2} = 64 \text{ Users}$$

This is the maximum amount of users the TRX is able to handle, in theory when taking into account the intra-cell interference. In reality the neighbouring cells produce inter-cell interference. If assumed that the inter-cell interference is the same as intra-cell interference then the user amount is split to 32 ($64/2 = 32$).

The E_b/N_{O} relationship is the point of interest. is a constant-like numeral relationship which may have several values, which later are related to radio interface bearer bit rates. Thus it can be stated that the E_b/N_{O} relationship has a remarkable effect on the TRX/cell capacity as far as simultaneous user amount is concerned. In the E_b/N_{O} relationship, the following issues should be considered:

- N_{O} is a local constant value type, which also contains some receiver specific values.
- E_b is a changing value in nature and its dependencies are described in the following bullets.
- Processing gain/spreading factor: the bigger the spreading factor value, the smaller the E_b.

- The higher base-band bit rate used, the bigger the E_b will be (this is a direct consequence from the previous point).
- Distance between terminal and BS receiver: the longer distance, the bigger E_b.
- Terminal motion speed: the higher speed used, the bigger E_b.

The calculation above is a rough way to estimate TRX capacity, there are many other ways to do this. For further details, refer to the *WCDMA for UMTS* by Holma and Toskala (2001).

4.3 Radio Network Controller

The RNC is the switching and controlling element of the UTRAN. The RNC is located between the Iub and Iu interface. It also has the third interface called Iur for inter-RNS connections. The implementation of the RNC is vendor dependent but some generic points can be highlighted (Figure 4.7).

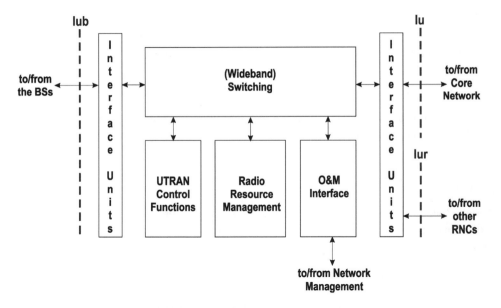

Figure 4.7.   RNC logical structure

As explained earlier, the RNC sees the BS as two entities being common transport and collection of Node B Communication contexts. The RNC controlling these for a BS is called *Controlling RNC* (CRNC).

Referring to the bearers, the RNC is a switching point between the Iu bearer and RB(s). One radio connection between the UE and the RNC, carrying user data is a RB. The RB in turn is related to the UE Context. UE Context is a set of definitions required in Iub in order to arrange both common and dedicated connections between the UE and the RNC. Since UTRAN utilises macrodiversity, the UE may have several RBs between itself and the RNC. This situation is known as soft handover and is discussed later on in this chapter. The RNC holding the Iu bearer for certain UE is called *Serving RNC* (SRNC).

The third logical role the RNC may have is *Drifting RNC* (DRNC). When in this mode, the RNC allocates UE Context through itself, the request to perform this activity comes from the SRNC through the Iur interface.

Both SRNC and DRNC roles are functionalities, which may change their physical location. When the UE moves in the network performing soft handovers, the radio connection of the UE will be accessed entirely through a different RNC than the SRNC, which originally performed the first RB set-up for this UE. In this case the SRNC functionality will be transferred to the RNC, which practically handles the radio connection of the UE. This procedure is called SRNC or SRNS Relocation.

The whole functionality of RNC can be classified into two parts, UTRAN Radio Resource Management (RRM) and control functions. The RRM is a collection of algorithms used to guarantee the stability of the radio path and the QoS of radio connection by efficient sharing and managing of the radio resources. The UTRAN control functions include all of these functions related to set-up, maintenance and release of the RBs including the support functions for the RRM algorithms.

4.3.1 Radio Resource Management

As explained in Chapter 1, the RRM is a management responsibility solely taken care of by UTRAN. RRM is located in both UE and RNC inside UTRAN. RRM contains various algorithms, which aim to stabilise the radio path enabling it to fulfil the QoS criteria set by the service using the radio path (Figure 4.8).

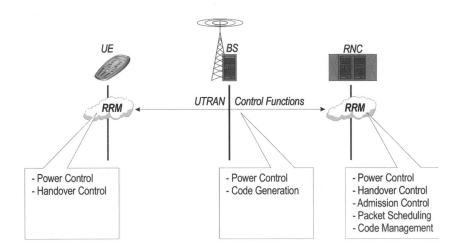

Figure 4.8.   Radio Resource Management and Radio Resource Control

The RRM algorithms must deliver information over the radio path, which is named UTRA Service. The control protocol used for this purpose is the Radio Resource Control (RRC) protocol. The UTRAN control functionality's are discussed later in this chapter. The RRC protocol is overviewed in Chapter 9.

The RRM algorithms to be shortly presented here are:

- Handover Control
- Power Control
- Admission Control (AC) and Packet Scheduling
- Code Management

4.3.1.1 Handover Control

Handover is one of the essential means to guarantee the user mobility in a mobile communications network, where the subscriber can move around. Maintaining the traffic connection
with a moving subscriber is made possible with the help of the *handover* function. The basic
concept is simple: when the subscriber moves from the coverage area of one cell to another, a
new connection with the target cell has to be set up and the connection with the old cell may
be released. Controlling the handover mechanism is however, quite complicated issue in
cellular systems and especially the Code Division Multiple Access (CDMA) system adds
some interesting ingredients to it.

Reasons Behind the Handover

There are many reasons why handover procedures may be activated. The basic reason behind
a handover is that the air interface connection does not fulfil the desired criteria set for it
anymore and thus either the UE or the UTRAN initiates actions in order to improve the
connection. In WCDMA, handover on-the-fly is used in context of circuit switched calls. In
the case of packet switched calls, the handovers are mainly achieved when neither the
network nor the UE has any packet transfer activity.

Regardless of the sort of handovers, they have the common nominator, handover criteria
that is why the handover should be performed. The logic behind how the need for the handover is investigated is also quite common. The handover execution criteria depend mainly
upon the handover strategy implemented in the system. However, most criteria behind the
handover activating the rest on the signal quality, user mobility, traffic distribution, bandwidth, and so forth.

Signal Quality Reason handover occurs when the quality or the strength of the radio signal
falls below certain parameters specified in the RNC. The deterioration of the signal is detected
by the constant signal measurements carried out by both the UE and the BS. The signal quality
reason handover may be applied both for the uplink and the downlink radio links.

Traffic reason handover occurs when the traffic capacity of a cell has reached its maximum
or is approaching it. In such a case, the UE nears the edges of the cell with high load may be
handed over to neighbouring cells with less traffic load. By using this sort of handover the
system load can be distributed more uniformly and the needed coverage and capacity can be
adapted efficiently to meet the demanded traffic within the network. Traffic reason handovers
may rely on pre-emption or directed retry approaches.

The number of handovers is straightforwardly dependent on the degree of UE mobility. If we
assume that the UE keeps on moving in the same direction then it can be said that the faster the
UE is moving the more handovers it causes to the UTRAN. To avoid undesirable handovers the
UE with high motion speed may be handed over, for instance, from micro cells to macro cells.

In case of the UE moving slowly or at all, on the other hand, it can be handed over from macro cells to micro cells to improve the radio signal strength and avoid consuming its battery.

The decision to perform a handover is always made by the RNC, that is currently serving the subscriber, except for the handover for traffic reasons. In the latter case the Mobile Switching Centre (MSC) may also make the decision. In addition to the above, many other reasons, for instance, change of services can also be a basis for the handover execution.

Handover Process

Figure 4.9 illustrates, a basic handover process consists of three main phases, including measurement phase, decision phase, and execution phase. The overall handover process discussed here is specifically related to the WCDMA system. Nevertheless, and as far as the basic principles are concerned, they are valid for any kind of cellular systems.

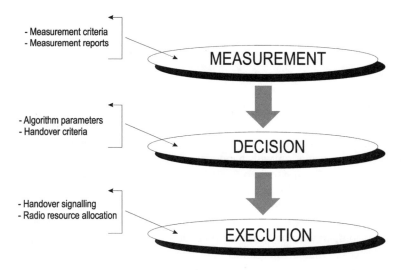

Figure 4.9.   Handover process

Handover measurement provision is a pivotal task from the system performance standpoint, this is because of two reasons. Firstly, the signal strength of the radio channel may vary drastically due to the fading and signal path loss, resulting from the cell environment and user mobility. Secondly, an excess of measurement reports by UE or handover execution by network increases the overall signalling, which is not desired.

For the handover purposes and during the connection the UE continuously measures the signal strength concerning the neighbouring cells, and reports the results to the network, to the radio access controller like RNC in WCDMA system.

According to the 3GPP Technical Specification (TS) 25.331, the UE measurements may be grouped into different categories, depending on what the UE should measure. The different types of measurements are:

- Intra-frequency measurements include measurements on the strength of the downlink physical channels for signals with same frequencies.

- Inter-frequency measurements include measurements on the strength of the downlink physical channels for signals with different frequencies.
- Inter-system measurements covers measurements on the strength of the downlink physical channels belonging to another radio access system than UTRAN, for example GSM.
- Traffic volume measurements contain measurements of the uplink traffic volume.
- Quality measurements include measurements of quality parameters, for example the downlink transport block error rate.
- Internal measurements include measurements of UE transmission power and UE received signal level.

The measurement events, on the other hand, may be triggered based on the following criteria:

- Change of best cell
- Changes in the Primary Common Pilot Channel (CPICH) signal level
- Changes in the P-CCPCH signal level
- Changes in the Signal-to-Interference Ratio (SIR) level
- Changes in the Interference Signal Code Power (ISCP) level
- Periodical reporting
- Time-to-trigger

Therefore, the WCDMA specification provides various measurement criteria to support the handover mechanisms in the system. In order to improve the system performance it is pivotal to select the most appropriate measurement procedure and measurement criteria as well as filtering intervals in association with handover mechanisms. The handover signalling load can be optimised by fine-tuning trade-off between handover criteria, handover measurements, and the utilised traffic model in the network planning.

Decision phase consists of assessment of the overall QoS of the connection and comparing it with the requested QoS attributes and estimates from neighbouring cells. Depending on the outcome of this comparison, the handover procedure may or may not be triggered.

The SRNC checks whether the values indicated in the measurement reports trigger any criteria set. If they trigger, then it allows executing the handover.

As far as the handover decision-making is concerned, there are two main types of handover:

- Network Evaluated Handover (NEHO)
- Mobile Evaluated Handover (MEHO)

In the case of a NEHO procedure, the network SRNC makes the handover decision, while in the MEHO approach the UE mainly prepares the handover decision. In the case of combined NEHO and MEHO handover types, the decision is made jointly by the SRNC and the UE.

It should be noted that even in a MEHO handover the final decision about the handover execution is made by the SRNC. The reason is that the RNC is responsible for the overall RRM of the system, and thus it is aware of the system overall load and other necessary information needed for handover execution.

Handover decision-making is based on the measurements reported by the UE and the BS as well as the criteria set by the handover algorithm. The handover algorithms, as such, are not subject to standardisation; rather they are implementation-dependent aspects of the system.

Therefore, advanced handover algorithms may be utilised freely, based on the available parameters in association with the network element measurement capabilities, traffic distribution, network planning, network infrastructure, and overall traffic strategy applied by operators.

The general principles of a handover algorithm are presented in Figure 4.10. In this example it is assumed that the decision-making criteria of the algorithm are based on the pilot signal strength reported by the UE. The following terms and parameters are used in this handover algorithm example:

- *Upper threshold*: is the level at which the signal strength of the connection is at the maximum acceptable level in respect with the requested QoS.
- *Lower threshold*: is the level at which the signal strength of the connection is at the minimum acceptable level to satisfied the required QoS. Thus the signal strength of the connection should not fall below it.
- *Handover margin*: is a pre-defined parameter, which is set at the point where the signal strength of the neighbouring cell (B) has started to exceed the signal strength of current cell (A) by a certain amount and/or for a certain time.
- *Active Set*: is a list of signal branches (cells) through which the UE has simultaneously connection to the UTRAN.

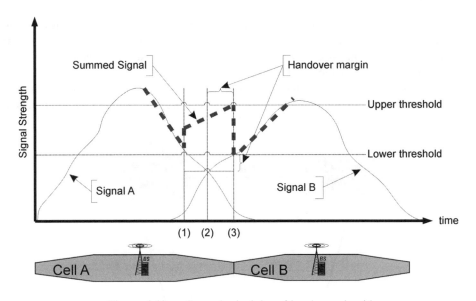

Figure 4.10.   General principles of handover algorithms

Assume that a UE camping in cell A is moving towards cell B. As the UE is moving towards cell B the pilot signal A – to which the UE currently has a connection – deteriorates, approaching to the lower threshold as shown in Figure 4.10. This may result in handover triggering during the following steps which can be distinguished:

1. The strength of the signal A becomes equal to the defined lower threshold. On the other hand, based on the UE measurements the RNC recognises that there is already a neighbouring signal (signal B in Figure 4.10) available, which has adequate strength for

improving the quality of the connection. Therefore, it adds the signal B to the Active Set. Upon this event, the UE has two simultaneous connections to the UTRAN and hence it benefits from the summed signal, which consists of signal A and signal B.

2. At this point the quality of signal B starts to become better than signal A. Therefore, the RNC keeps this point as the starting point for the handover margin calculation.

3. The strength of signal B becomes equal or better than the defined lower threshold. Thus its strength is adequate to satisfy the required QoS of the connection. On the other hand, the strength of the summed signal exceeds the defined upper threshold, causing additional interference to the system. As a result, the RNC deletes signal A from the Active Set.

It should be noted that the size of the Active Set may vary but usually it ranges from 1 to 3 signals. In this example the size of Active Set is 2 between event (1) and (3).

Because the direction of UE motion varies randomly it is possible that it comes back towards the cell A instantly after the first handover. This results to a so-called *ping- pong* effect, which is harmful for the system in terms of capacity and overall performance. Using the handover margin or hysteresis parameter is to avoid the undesired handovers, which cause additional signalling load to the UTRAN.

Handover Types

Depending on the diversity used in association with handover mechanisms, they can be categorised as hard handover, soft handover, and softer handover. The hard handover can be further divided into intra-frequency and inter-frequency hard handovers. All of these mechanisms are provided by the UMTS system.

During the handover process, if the old connection is released before making the new connection it is called a *hard handover*. Therefore, there are not only lack of simultaneous

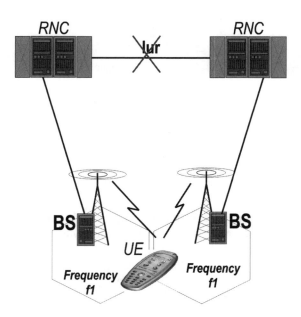

Figure 4.11.   Intra-frequency hard handover

signals but also a very short cut in the connection, which is not distinguishable for the mobile user.

In case of *inter-frequency* hard handover the carrier frequency of the new radio access is different from the old carrier frequency to which the UE was connected. On the other hand, if the new carrier, to which the UE is accessed after the handover procedure is the same as the original carrier then there is an *intra-frequency* handover in question.

Figures 4.11 and 4.12 show hard handover situations when the neighbouring BSs may transmit with same frequencies or with the different frequency, respectively.

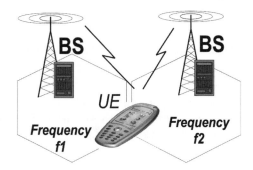

Figure 4.12.   Inter-frequency hard handover

In Figure 4.11, the neighbouring RNC is not connected by Iur interface due to the radio network planning strategy or transmission reasons and hence the inter-RNC soft handover is not possible. Under these circumstances, intra-frequency hard handover is the only handover to support the seamless radio access connection and subscriber mobility from the old BS to the new BS. In fact, this leads to an inter RNC handover event, in which the MSC is also involved.

Generally, the frequency reuse factor is one for WCDMA, meaning that all BS's transmit on the same frequency and also all UE share a common frequency within the network. This does not mean, however, that the frequency reusing cannot be utilised in WCDMA at all. Therefore, if different carriers are allocated to the cells for some other reason, inter-frequency handover is required to ensure handover path from one cell to another cell in the cell cluster.

Inter-frequency handover also occurs in Hierarchical Cell Structure (HCS) network between separate cell layers, for instance, between macro cells and micro cells, which use different carrier frequencies within the same coverage area. In this case the inter-frequency handover is used not only because the UE would otherwise lose its connection to the network, but also in order to increase the system performance in terms of capacity and QoS. Inter-frequency hard handover is always a NEHO.

Moreover, inter-frequency handover may happen between two different radio access networks (RAN's), for instance, between GSM and WCDMA. In this context it can also be called inter-system handover (Figure 4.13). An inter-system handover is always a type of inter-frequency, since different frequencies are used in different systems.

The possibility to perform an inter-system handover is enabled in the WCDMA by a special functioning mode, *Compressed Mode (Slotted Mode)*. Compressed Mode is also known with name Slotted Mode. From the WCDMA point of view and when the UE is in

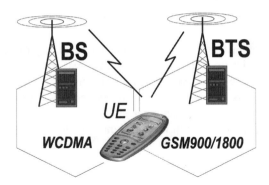

Figure 4.13.   Inter-system hard handover

Slotted Mode, the spreading factor value of the channel can be reduced. As a consequence, the radio interface connection uses only part of the space in the WCDMA frame slot. The rest of the slot can be used for other purposes by the UE; for instance, the UE may measure surrounding GSM cells. In other words, this mechanism is the way to implement GSM/UMTS interoperability requirement in UTRAN. In addition, the Slotted Mode can be achieved by reducing the data rate using the higher layer controlling and also by reducing the symbol rate in association with the physical layer multiplexing. When the UE uses Uu interface in this mode, the contents of the WCDMA frame is "compressed" a bit in order to open a time window through which the UE is able to peek and decode the GSM Broadcast Control Channel (BCCH) information. Additionally, both the WCDMA RAN and GSM Base Station System (BSS) must be able to send each other's identity information on the BCCH's so that the UE is able to perform the decoding properly.

The inter-system WCDMA and GSM handover can be applied in areas where WCDMA and GSM systems co-exist. Inter-system handover is required to complement the coverage areas of each other in order to ensure continuity of services. The inter-system handover can also be used to control the load between GSM and WCDMA systems, when the coverage area

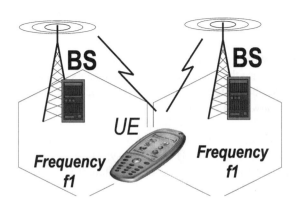

Figure 4.14.   Intra-frequency soft handover

of the two systems overlap with each other. Other inter-system handover reasons might be the service requested by the UE and user's subscription profile.

Inter-system handover is a NEHO. However, also the UE must support inter-system handover completely. The RNC recognises the possibility of inter-system handover based on the configuration of the radio network mainly based on neighbour cell definitions and other control parameters. The same applies for the BSC in the GSM RAN side.

Unlike in hard handover, if a new connection is established before the old connection is released then the handover is called *soft handover*. In the WCDMA system, the majority of handovers are intra-frequency soft handovers. As is illustrated in Figure 4.14, in soft handover the neighbouring BS involved in the handover event transmits the same frequency.

Soft handover is performed between two cells belonging to different BS's but not necessarily to the same RNC. In any case the RNC involved in the soft handover must co-ordinate the execution of the soft handover over the Iur interface. In a soft handover event the source and target cells have the same frequency. In case of a circuit switched call, the terminal is actually performing soft handovers almost all the time if the radio network environment has small cells. There are several variations of soft handover, including softer and soft-softer handovers.

A softer handover is a handover by which a new signal is either added to or deleted from the Active Set, or replaced by a stronger signal within the different sectors, which are under the same BS (Figure 4.15).

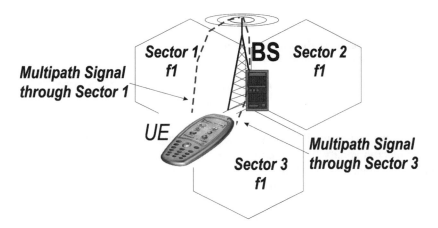

Figure 4.15. Intra-frequency softer handover

In softer handover, the BS transmits through one sector but receives from more than one sector. In this case the UE has active uplink radio connections with the network through more than one sector populating the same BS.

When soft and softer handovers occur simultaneously, the term soft-softer handover is usually used. A soft-softer handover may occur, for instance, in association with inter-RNC handover, while an inter-sector signal is added to the UE's Active Set along with adding a new signal via another cell controlled by another RNC.

Referring to the soft handover and Active Set, there are two terms describing the handling of the multipath components. These are microdiversity and macrodiversity.

Microdiversity means the situation where the propagating multipath components are combined in the BS as shown in Figure 4.16. The WCDMA utilises multipath propagation. This means that the BS RAKE receiver, which was already mentioned in 4.1.1.3, is able to determine, differentiate and sum up several signals received from the radio path. In reality, a signal sent to the radio path is reflected from the ground, water, buildings, etc. and at the receiving end the sent signal can be "seen" as many copies, all of them coming to the receiver at a slightly different phase and time. The microdiversity functionality at the BS level combines different signal paths received from one cell, and in the case of sectored BS, the outcome from different sectors, which is also referred to as softer handover.

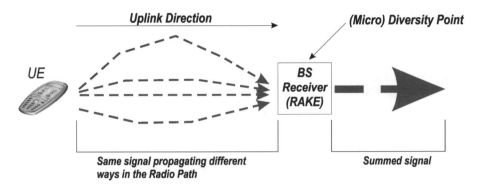

Figure 4.16.   Micro diversity in BS

Because of the fact that the UE may use cells belonging to different BS's or even different RNC's the macrodiversity functionality also exists on the RNC level. However, the way of combining the signals is quite different than in the microdiversity case at BS because there is no RAKE receiver at RNC. Therefore, other approaches like the quality of data flow may be utilised to combine or select the desired data stream. Figure 4.17 presents a case in which the UE has a three-cell Active Set in use and one of those cells is connected to another RNC. In this case, the BS's firstly sum up the signal concerning the radio paths of their own and final summing of the data stream is done on the RNC level.

Concerning soft and softer handovers, the idea is that the subjective call quality will be better when the "final" signal is constructed from several sources (multipath). In GSM the subjective call quality depends on the transmission power used: it can be roughly stated that more power, the better quality. In WCDMA, the terminals cannot use a lot of power because transmission levels that are too high will start blocking the other users, thus, the better way to gain better subjective call quality is to utilise multipath propagation.

As a conclusion it can be stated that soft and softer handovers consume radio access capacity because the UE is occupying more than one radio link connection in the Uu interface. On the

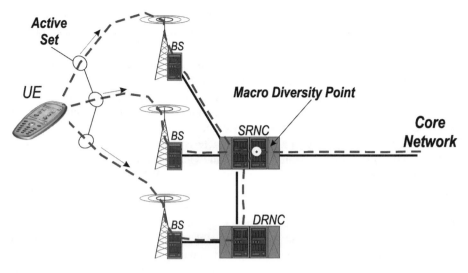

Figure 4.17.   Macro diversity in RNC

other hand, the added capacity gained from the interference reduction is bigger and hence the system capacity is actually increased if soft and softer handovers are reasonably used.

From system structural architecture standpoint, UMTS network supports the following types of handovers:

- Intra BS/inter-cell handover (softer handover)
- Inter BS handover, including hard and soft handovers
- Inter RNC handover, including hard, soft, and soft-softer handovers
- Inter MSC handover
- Inter Serving GPRS Support Node (SGSN) handover
- Inter system handover

4.3.1.2 Power Control

Power control is an essential feature of any CDMA based cellular system. Without utilising an accurate power control mechanism, these systems do not operate. In the following subsections we first describe why power control is so essential for these cellular systems and what are the main factors behind this fact. We then describe what kind of power control mechanisms are utilised in the WCDMA-FDD radio access.

The main reasons for implementing power control are the near–far problem, interference dependent capacity of the WCDMA and the limited power source of the UE. Unlike Frequency Division Multiple Access (FDMA) and Time Division Multiple Access (TDMA), which are bandwidth-limited multiple accesses, WCDMA is an interference-limited multiple access. In FDMA and TDMA, power control is applied to reduce inter-cell interference within the cellular system that arises from frequency reuse while in WCDMA systems, the purpose of the power control is mainly to reduce the intra-cell interference. Meeting these targets requires optimisation of the radio transmission power, meaning that the power of every transmitter is adjusted to the level required to meet the requested QoS. Determining the transmission power level is, however, a very sophisticated task due to dynamic variation of the radio channel.

Whatever the radio environment is, the received power should be at an acceptable level, for example at the BS for the uplink to support the requested QoS. The target of power control is to adjust the power to the desired level without any unnecessary increase in the UE transmit power. It takes care that, transmit power is just within the required level and neither higher nor less, taking into account the existing interference in the system.

The influence of multipath propagation characteristics and technical characteristics of WCDMA system, for instance, simultaneous bandwidth sharing and near–far phenomena result in the fact that power control, is essential for the WCDMA system, to overcome drawbacks caused by the radio environment and the nature of the electromagnetic wave. Without power control, phenomena like fading and interference, drive down the system stability and ultimately degrade its performance dramatically.

Maximising system capacity is an invaluable asset for both advanced cellular technology suppliers and cellular network operators. System capacity is maximised if the transmitted power of each terminal is controlled so that its signal arrives at the BS with the minimum required SIR. If a terminal's signal arrives at the BS with too low a received power value then the required QoS for the radio connection can not be met. If the received power value is too high, the performance of this terminal is good, however, interference to all the other terminal transmitters sharing the channel is increased and may result in unacceptable performance for other users, unless their number is reduced.

Due to the fact that in the WCDMA system the total bandwidth is shared simultaneously, other users can be experienced as a noise–like interference for a specific user. In case the power control mechanism is missing or operates imperfectly, common sharing of the bandwidth creates a severe problem, referred to as the near-far effect. In near-far situations the signal of the terminal that is close to the serving BS may dominate the signal of those terminals, which are far away from the same BS. Figure 4.18 illustrates the situation in

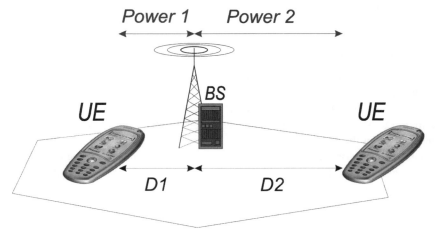

D1, D2 = Distances between UE and BS

Figure 4.18.   Near-far effect in the CDMA system

which the near-far problem may occur. The main factors that cause the near-far problem result from the path-loss variation of simultaneous users with different distances from the BS, fading variation, and other signal-power variation of the users caused by the radio wave propagation mechanisms described in Chapter 3.

In WCDMA, the near-far effect can be mitigated by applying power control mechanism, diversity techniques, soft handover, multi-user receiver and generally near-far resistance receivers. Because of the crucial drawback of the near-far effect on the performance of WCDMA system, its mitigating is one of the pivotal purposes of power control mechanisms. These mechanisms have considerable impact on the WCDMA system capacity.

Basic approaches to power control

Due to the facts described previously, it is relatively easy to determine that in the uplink case the optimal situation from the BS receiver point of view is that the power representing one UE's signal is always equal when compared to the other UE signals regardless of their distance from the BS. If so, the SIR will be optimal and the BS receiver is able to decode the maximal number of transmissions. In reality, however, the radio channel is extremely unstable and also the requested radio services vary for different users and even for the same user and during one radio connection. Therefore the transmission power of UE should be controlled very accurately by utilising efficient mechanisms.

To achieve this, power control has been well investigated and many power control algorithms have been developed since the origin of the CDMA scheme. These include distributed, centralised, synchronous, asynchronous, iterative, and non-iterative. Most of the existing algorithms utilise either SIR or transmit power as a reference point in the power control decision-making process.

The primary principle of Centralised Power Control (CPC) schemes is that they keep the overall power control mechanism centralised. As a result they require a central controller, which should have knowledge of all the radio connections in the RAN.

In contrary to CPC methods, distributed power control methods do not utilise central controller. Instead they distribute the controlling mechanism within the RAN and toward the edge of it. This feature makes them of special interest. The CPC approaches bring about added complexity, latency and network vulnerability. The main advantage of distributed power control algorithm is that it can respond more adaptively to variable QoS, which is vastly important for cellular systems with packet-based transmission characteristics like WCDMA.

Power Control Mechanism in UTRAN (WCDMA-FDD)

In WCDMA, power control is employed in both the uplink and the downlink. Downlink power control is basically for minimising the interference to other cells and compensating for other cells interference as well as achieving acceptable SIR. However, for downlink, power control is not as vital as it is for the up-link transmit power adjustment. It is still implemented also for the downlink, because it improves system performance by controlling interference to other cells.

The main target of the uplink power control is to mitigate the near-far problem by making the transmission power level received from all terminals as equal as possible at the home cell for the same QoS. Therefore uplink power control is for fine-tuning of terminal transmission power, resulting in the mitigation of the intra-cell interference and near-far effect. It should be mentioned that the power control mechanism specified for the WCDMA is, in principle, a distributed approach.

The power control mechanisms used in the GSM are clearly inadequate to guarantee this situation in WCDMA and thus the WCDMA has a different approach to the matter. In GSM, the power control is applied for the connection once or twice per second, but due to its critical nature in WCDMA, the power used in the connection is adjusted 1500 times per second, that is the power control cycle is repeated in association with each radio frame in association with DCH. Therefore, the power adjustment steps are considerably faster than in GSM.

To manage the power control properly in WCDMA, the system uses two different defined power control mechanisms as shown in Figure 4.19. These power control mechanisms are:

- Open Loop Power Control (OLPC)
- Closed Loop Power Control (CLPC), including inner and outer Loop Power Control mechanisms

By applying all of these different power control mechanisms together, the UTRAN benefits from the advantages of CPC as well by overlaying the inner closed loop control with the outer closed loop control mechanism in order to keep the target SIR in an acceptable level.

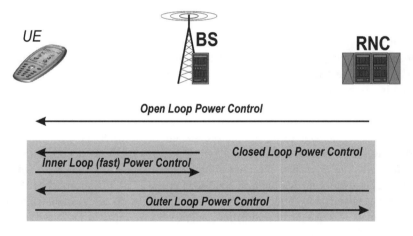

Figure 4.19.   WCDMA power control mechanisms

Open Loop Power Control (OLPC)

In the OLPC, which is basically used for uplink power adjusting, the UE adjusts its transmission power based on an estimate of the received signal level from the BS CPICH when the UE is in idle mode and prior to Physical Random Access Channel (PRACH) transmission. In addition to that, the UE receives information about the allowed power parameters from the cell BCCH when in idle mode. The UE evaluates the path loss occurring and based on this difference together with figures received from the BCCH and the UE it is able to estimate what might be an appropriate power level to initialise the connection.

Figure 4.20 illustrates the OLPC as applied for the uplink case. In this process, the UE estimates the transmission signal strength by measuring the received power level of the pilot

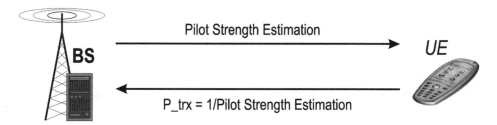

Figure 4.20. OLPC for uplink

signal from the BS in the downlink, and adjusts its transmission power level in a way that is inversely proportional to the pilot signal power level. Consequently, the stronger the received pilot signal, the lower the UE transmitted power.

In the case of Frequency Division Duplex (FDD)-based WCDMA, OLPC alone is not adequate for adjusting the UE transmission power, because fading characteristics of the radio channel vary rapidly and independently for the uplink and the downlink. Therefore, in order to compensate the rapid changes in the signal strength CLPC mechanism is also needed. Nevertheless, OLPC is useful for determining the initial value of transmitted power and mitigating drawbacks of the log-norm-distributed path-loss and shadowing.

Closed Loop Power Control (CLPC)

CLPC is utilised for adjusting the transmission power when the radio connection has already been established. Its main target is to compensate the effect of rapid changes in the radio signal strength and hence it should be fast enough to respond to those changes.

Figure 4.21 illustrates the basic uplink CLPC mechanism specified for WCDMA. In this case the BS commands the UE to either increase or decrease its transmission power with a cycle of 1.5 kHz (1500 times per second) by 1, 2, or 3 dB step-sizes. The decision whether to increase or decrease the power is based on the received SIR estimated by the BS. When the BS receives the UE signal it compares the signal strength with the pre-defined threshold value

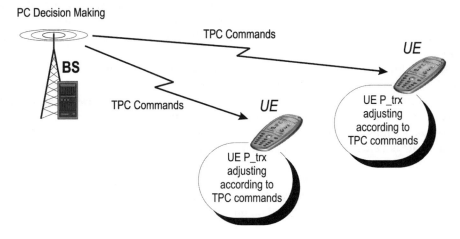

Figure 4.21. Basic uplink CLPC mechanism

at the BS. If the UE transmission power exceeds the threshold value, the BS sends a Transmission Power Command (TPC) to the UE to decrease its signal power. If the received signal is lower than the threshold target the BS sends a command to the UE to increase its transmission power. It should be emphasised that various measurement parameters, for example SIR, signal strength, Frame Error Ratio (FER) and Bit Error Ratio (BER) can be used for comparing the quality of the received power and making a decision for controlling transmission power.

It should be also noted that CLPC is utilised for adjusting transmission power in the downlink as well. In the case of the downlink CLPC, the roles of the BS and the UE are interchanged. That is, the UE compares the received signal strength from the BS with a predefined threshold and sends the TPC to the BS to adjust its transmission power accordingly.

In WCDMA, CLPC mechanism consists of Inner Loop and Outer Loop variants. What was described so far, was related to the Inner Loop variant, which is the fastest loop in WCDMA power control mechanism and hence it is occasionally referred to as fast power control.

Another variant of the CLPC is the Outer Loop Power Control (OLPC) mechanism. The main target of OLPC is to keep the target SIR for the uplink Inner Loop Power Control mechanism in an appreciated quality level. Thanks to the macrodiversity, the RNC is aware of the current radio connection conditions and quality. Therefore, the RNC is able to define the allowed power levels of the cell and target SIR to be used by the BS when determining the TPCs. In order to maintain the quality of the radio connection, the RNC uses this power control method to adjust the target SIR and keep the variation of the quality of the connection in control. By doing this, the network is able to compensate changes in the radio interface propagation conditions and to achieve the maximum target quality for the connection BER and the FER observation. In fact, the OLPC fine-tunes the performance of the Inner Loop Power Control.

Together, the Open and CLPC mechanisms have considerable impact on in the terminal's battery-life and overall system capacity in any cellular system and specifically in CDMA based mobile systems.

Power Control in Specific Cases

In addition to ordinary power control mechanisms used in WCDMA, there are additional approaches to cope with specific cases. These include controlling transmission power associated with soft handover, Site Selection Diversity (SSDT), and compressed mode (Slotted Mode).

In soft handover state, the transmission power of the UE is adjusted based on the selection of the most suitable power control command from those TPC's that it receives from different BS's to which it has simultaneous radio links (Figure 4.22). In this case because the UE receives more than one TPC command from the different BS's independently, the received TPC commands may differ from each other. This may result from the fact that power control commands are not efficiently protected against errors or simply it may result from the network environment. This leads to a conflict situation for the UE. The basic approach to resolve this problem is that, if at least one of the TPC commands refers to the decrease in transmission power then the UE decreases its power. It is also defined that the UE can utilise a threshold for detecting the reliable commands, based on this it can decide whether to increase or decrease its transmission power.

SSDT is another special power control solution. The principle of SSDT is that the BS with

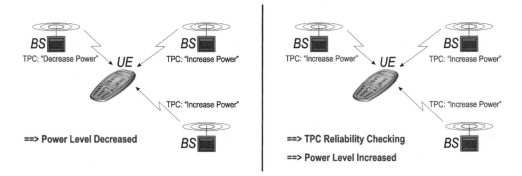

Figure 4.22.   Power control associated to soft handover

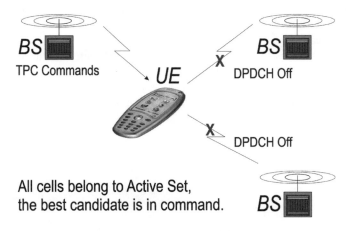

Figure 4.23.   Site Selection Diversity (SSDT)

the strongest signal is dynamically chosen as the only transmitting BS (Figure 4.23). Then, the other BS's, to which the UE has simultaneously radio connected turn off their Dedicated Physical Data Channels (DPDCH). Therefore, the transmit power is adjusted based on the power control commands of the BS with the strongest signal. It seems that this method reduces the downlink interference generated while the UE is in soft handover.

Yet in compressed mode, the transmission and reception of the BS and UE are ceased for a predefined period to make room for performing the inter frequency radio measurements, for instance in conjunction with inter system handover event. As a result there is also a break in the transmission power adjusting mechanism. In this case the receiver of the power control commands, for instance, the UE for the uplink case is allowed to increase or decrease it's transmit power with larger step-sizes to reach the desirable SIR level as fast as possible.

4.3.1.3 Admission Control and Packet Scheduler

WCDMA radio access has several limiting factors, some of them being absolute and others environment dependent. The most important, and at the same time the most difficult to control is the interference occurring in the radio path. Due to the nature and basic characteristics of WCDMA, every UE accessing the network generates a signal and simultaneously this signal can be interpreted to be interference, from the other UE point of view. When the WCDMA cellular network is planned, one of the basic criteria for planning is to define the acceptable interference level with which the network is expected to function correctly. This planning-based value and the actual signals the UE transmit set practical limits for the radio interface capacity.

To be more specific, a defined SIR value is used in this context. Based on radio network planning the network is, in theory stable, as long as the defined SIR level is not exceeded within the cell. Practically, it means that in the BS receiver, the interference and the signal must have a certain level of power difference in order to extract one signal (code) out from the other signals using the same carrier. If the power distance between interfering components and the signal is too small the BS is unable to extract an individual signal (code) out from the carrier any more. Every UE having a bearer active through the cell "consumes" a part of the SIR and the cell is used up to its maximum level when the BS receiver is unable to extract the signal(s) from the carrier.

The main task of AC is to estimate whether a new call can have access to the system without sacrificing the bearer requirements of existing calls (Figure 4.24). Thus the AC algorithm should predict the load of the cell if the new call is admitted. It should be noted that the availability of the terrestrial transmission resources is verified, too. Based on the AC, the RNC either grants or rejects access.

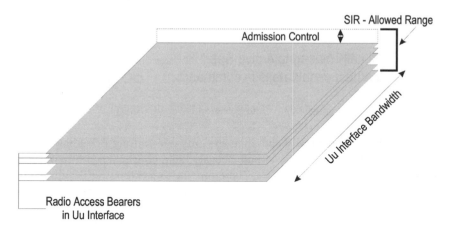

Figure 4.24.   Admission Control (AC)

Theoretically, it can be found out that the SIR or Interference Margin has direct relationship with the Cell Load. If we express the Cell Load with Load Factor as a parameter, which expresses the cell percentual load as shown in Figure 4.25 and mark the Interference Margin with I, it leads to the following equation:

$$I = 10 \times \text{Log}\left(\frac{1}{1 - \text{Load Factor}}\right)$$

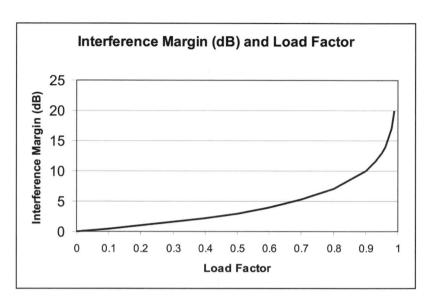

Figure 4.25.   Interference margin as function of cell load

When placing together the Interference Margins calculated with different Load Factor values we arrive at the results shown in Figure 4.25.

Based on the graph it is fairly easy to indicate that when the Cell Load exceeds 70%, the interference in that cell will be very difficult to control. This is why the WCDMA radio network is normally dimensioned with expected capacity equivalent to Load Factor value 0.5 (50%), this value has a safety margin in it and the network will most likely operate steadily.

Due to the nature of traffic in the UMTS system both Real Time (RT) and Non-Real Time (NRT) traffic should be controlled and balanced carefully. The ways of controlling the RT and NRT traffic are distinctly different. The main difference is that for a RT (circuit call) service a RAB is maintained continuously in a dedicated state. In case of a NRT service (packet connection), the channel state varies as the consequence of the RB bit rate variation as well as the system load. So, the bit rate variation and the bursting characteristic of packet connection should be taken into account during packet connection admission process.

Figure 4.26 illustrates the main characteristics of a packet connection based on a World Wide Web (WWW) browsing session. During a WWW packet connection the user typically sends a WWW address to the network for data fetching purposes. As a response, the network downloads data, typically an HyperText Markup Language (HTML)–page , from the desired address. The desired address may also lead to document/file download. At the next step, a normal user is consuming a certain amount of time for studying the information, referring to it as the reading time. During the reading time there is no need for the AC to keep the RAB in a dedicated state, but rather the radio resources should be used for other purposes, e.g. other incoming and outgoing circuit or packet switched connections. The overall packet connection information (session) however, is maintained in upper level. So, if after the reading time the user wants to have another page downloaded, the pre-condition for the call is already available.

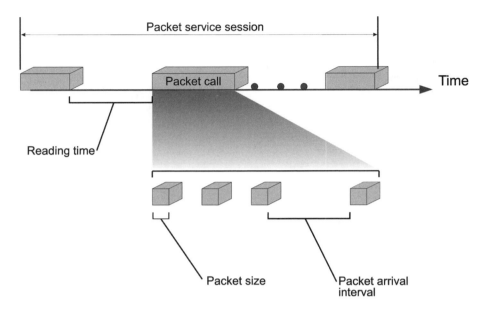

Figure 4.26.   Characteristic of a packet service session.

The AC is responsible for the handling of packet connections with bursty traffic, having a very random arrival time, number of packet call per session, reading time, as well as number of packets within a call. Therefore, the AC must utilise very sophisticated traffic models and statistical approaches in order to control optimally the requested RAB(s). By scheduling the NRT RABs, and accepting, queuing, or rejecting the RT RABs. It is the responsibility of the AC to keep and control the QoS of the accepted RABs and their influence to the overall performance of the UTRAN.

4.3.1.4 Code Management

Both channelisation and scrambling codes used in the Uu interface connections are managed by the RNC. In principle, the BS could manage them, but then the system may behave unstable when the RNC is otherwise controlling the radio resources in terms of, for example soft handovers. When the codes are managed by the RNC, it is also easier to allocate Iub data ports for multipath connections.

The Uu interface requires two kinds of codes for proper functionality. Part of the codes must correlate with each other to a certain extent. The others must be orthogonal or they do not correlate at all. Every cell uses one scrambling code; the UE is able to make separation between cells by recognising this code. Under every scrambling code the RNC has a set of channelisation codes. This set is the same under every scrambling code. The BCH information is coded with a scrambling code value and thus the UE must find the correct scrambling code value first in order to access the cell. When a connection between the UE and the network is established, the channels used must be separated. The channelisation codes are used for this purpose. The information sent over the Uu interface is spread with a spreading code per channel and the spreading code used is scrambling code $\times$ channelisation code.

4.3.2 UTRAN Control Functions

In order for the UTRAN to control and manage the RBs, which is essential to provide the RAB service, it should perform other functions in addition to the RRM algorithms. These can be classified as:

- System Information Broadcasting
- Random Access and Signalling Bearer Set-up
- RB Management
- UTRAN Security Functions
- UTRAN level Mobility Management (MM)
- Database Handling
- UE positioning

4.3.2.1 System Information Broadcasting

An important function of the RNC is to handle the system information task. System information is used to maintain both the radio connections between the UE and the UTRAN and also control the overall operation of the UTRAN. The RNC broadcasts the system information elements to assist the UTRAN controlling functions by providing the UE with the essential data needed when communicating with the UTRAN. For example, for providing radio measurement criteria, paging occasion indication, radio path information, assistance data for positioning purposes, etc. The system information can be received by the UE both in idle mode and all connected states when it has been identified by the UTRAN. System information services may also be used, for example by the CN to provide broadcast services. The RNC utilises point-to-multi-point system information broadcasting to keep the UE in touch with the UTRAN when necessary.

From protocol architecture point of view, the system information broadcast functionality is part of the RRC and it is terminated at the RNC.

Figure 4.27 illustrates the system information structure, which together with segmentation class information is used by the RNC as basis for controlling the system information segmentation and scheduling. As shown, the system information elements are organised based on the Master Information Block (MIB), two optional Scheduling Blocks (SB) and System Information Blocks (SIB), which contains actual system information. 3GPP specification TS 25.331 defines up to 17 different types of SIB. Some of those may also contain sub-SIBs. A MIB contains reference and scheduling information to a number of SIBs in a cell. It may also contain reference and scheduling information to one or two SBs, which in turn gives reference and scheduling information for additional SIBs. Therefore, only the MIB and SBs can contain scheduling information for a SIB.

Because SIBs are characterised differently and grouped-specifically in terms of their repetition rate and criticality, the RNC can use them for different purposes, for example it uses SIB#1 to inform the UE about timers and counters to be used in idle and connected mode. Other examples are SIB#2 and SIB#3 which the RNC uses to inform the UE about the UTRAN level MM and cell selection and re-selection, respectively.

The RNC may handle the constructing process together with the BS, for example in modifying the system information scheduling. Nevertheless the controlling function of

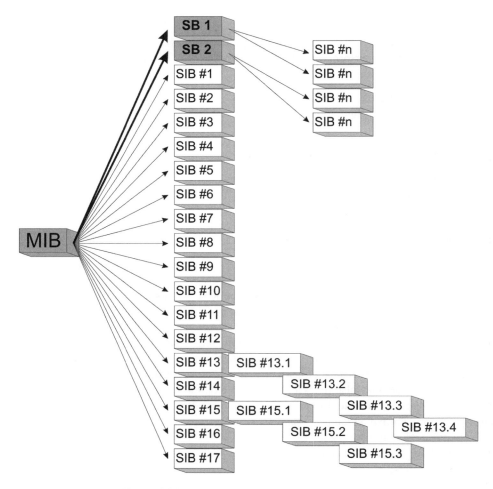

Figure 4.27.   Basic structure of system information

system information belongs entirely to the RNC. Once the RNC has structured the system information data it forwards the information to the BS to be broadcasted over air interface to the UE. Based on the air interface situation the BS informs the RNC about the ability to broadcast the system information on the Uu interface (Figure 4.28).

4.3.2.2 Initial Access and Signalling Connection Management

Before the RNC can map any requested RABs to the RB, it needs to create the signalling connection between the UE and the CN. In this context, 3GPP specification Technical Report (TR) 25.990 defines the *signalling connection* as an acknowledged-mode link between the UE and the CN to transfer higher layer information between the entities in the Non Access Stratum (NAS). In order to do this the RNC uses the RRC connection services in creating the *Signalling Radio Bearer* (SRB) between the UE and the UTRAN to provide the transferring services for the signalling connection. Once the RNC has created the SRB, it uses the first

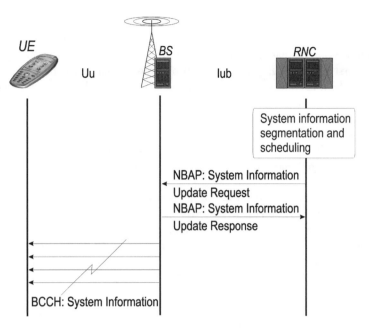

Figure 4.28.   System information broadcasting

RLC RBs to convert and transfer the signalling connection through the UTRAN to the UE. *RB* is defined as the services provided by the RLC layer for transfer of user data between the UE and the RNC. It is however, defined that the primary RBs are also used for signalling connection purposes (Figure 4.29).

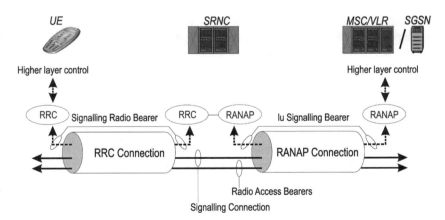

Figure 4.29.   Relation between the SRB, RB, signalling connection, RAB and the role of RNC

   The whole process starts when the UE enters the idle mode by turning on its power. After power-on, the UE attempts to make contact with the UTRAN. The UE looks for a suitable cell in the UTRAN and chooses the cell to provide available services, and tunes to its control channel. This choosing is known as "camping on the cell". The camping includes a cell search, where the UE searches for a cell and determines the downlink scrambling code and frame synchronisation of that cell. The cell search is down within different steps.

During the first step the UE uses the Synchronisation Channel (SCH)'s primary synchronisation code, which is common to all cells, to acquire slot synchronisation to a cell. During the second step, the UE uses the SCH's secondary synchronisation code to find frame synchronisation and identify the code group of the cell found in the first step. During the third and last step, the UE determines the exact primary scrambling code used by the cell found within the UTRAN. If the UE identifies the primary scrambling code then the P-CCPCH can be detected and the system and cell specific broadcast information defined by the RNC can be read.

In this regard the UE should be aware of the way the system information is structured and scheduled by the RNC based on the MIBs, SBs and SIBs. If the Public Land Mobile Network (PLMN) Identity found from the MIB on BCH matches the PLMN (or list of PLMNs) that the UE is searching for, it can continue to read the remaining BCH information, for example parameters for the configuration of the common physical channels in the cell (including PRACH and secondary CCPCH). Otherwise the UE may store the identity of the found PLMN for possible future use and restarts the cell search mechanism.

The first cell search for a PLMN is normally hardest to the UE, since it has to scan through a number of scrambling codes when trying to find the correct one. Once the UE obtains the necessary information to capture the BS controlled by the corresponding RNC it can request the *initial access* to the UTRAN, resulting in the transition from idle mode to the connected mode.

Until now the RNC was involved in the controlling function in terms of system information broadcasting and controlling the related BS to which UE was trying to get radio connection. Thereupon the RNC participates actively in controlling the access provisioning by considering the context of RRC connection set-up message requested by the UE via the Random Access Channel (RACH). In this association, the RNC has a central role to control the radio connection by checking the UE identity, the reason for the requested RRC connection and the UE capability, which helps it in allocating the initial signalling RB for the UE. Based on this information the RNC decides whether or not to allocate the signalling RB to the UE, which is used to carry the rest of the signalling for initial access and providing all the services after that. In case the RNC does not accept the access request then the UE can restart the initial access within a predefined time span.

No matter from which direction the higher layer service is requested, as is shown in the Figure 4.30, the RRC connection requested for creating the SRB is always started by the UE. When the RNC receives the RRC connection set-up request, it sets up a radio link over Iub interface to the BS to which the UE is going to have radio connection. In this context, the *Radio Link* is defined as a logical association between a single UE and a single UTRAN access point and its physical realisation comprises one or more RB transmissions. If this step is successful then the RNC informs the UE that the RRC connection is set-up and the UE responses to the RNC concerning the RRC connection completion.

Now the SRB is ready between the RNC and the UE and hence the RNC can convert and transfer the signalling connections and RABs between the UE and the CN. It should be mentioned that regardless of how many signalling connections and RABs exist between the UE and the CN, there is only one RRC connection used by the RNC to control and

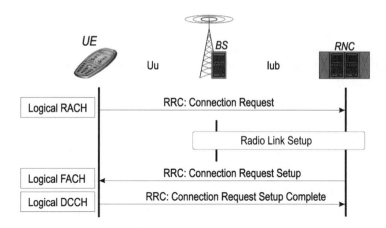

Figure 4.30.   RRC connection set-up

transfer them between the UE and UTRAN. It is the function of RNC to reconfigure the lower layer services needed, based on the amount of the signalling connections and RABs variation, which should be transferred.

4.3.2.3 Radio Bearer Management

Once the signalling Radio Bearer (RB) is established between the radio network and the UE as was previously described, it is time for the RNC to map the requested RAB(s) onto the RB for transferring between the UTRAN and the UE. In this way the RNC creates the illusion about a fixed bearer between the CN and the UE.

After the requested RABs have been negotiated between the UE and CN via the signalling connection it is forwarded to the RNC for further actions. In this regard the main function of the RNC is to analyse the attributes of the requested RAB(s), evaluate the radio resources needed for these, activate or reconfigure the radio channels by using the lower layer services and map the requested RAB(s) to the RB. Therefore, the RNC covers all of the control functions related to the supporting of the RAB(s) as well as configuring the available radio resources for that purpose.

As shown in Figure 4.29, the RAB is carried within the RRC connection between the RNC and UE over the radio interface and within the Radio Access Network Application Part (RANAP) protocol connection between RNC and CN over the Iu interface. In this association RNC acts as protocol converter between the RAN and CN. The RNC maps the requested RABs onto the RBs by using the current radio resource information and controls the lower layer services.

4.3.2.4 UTRAN Security

Radio connection security is another important function handled by the RNC. The RNC is involved in both integrity checking and ciphering mechanisms. The former is used for protecting the signalling connection between the UE and the UTRAN over the radio interface and the latter is used for protecting the user data transferred between the UE and the UTRAN.

The RNC ciphers the signalling and user data by using the defined integrity and ciphering algorithms. In this regard it needs to generate random numbers and maintain time-dependent counter values for the integrity checking of signalling messages. It also verifies and deciphers the received messages by using those algorithms. The algorithms and other UMTS security related issues are thoroughly addressed in Chapter 8.

4.3.2.5 UTRAN level Mobility Management

UTRAN level Mobility Management (MM) refers to those functions, which RNC handles in order to keep the UE in touch with the UTRAN radio cells, taking into account the user's mobility within UTRAN and the type of traffic or RAB(s) it is using.

As was mentioned in RRM subsection, the nature of traffic in the UMTS network environment is substantially different from the traditional circuit-switched type of traffic. This means that the UTRAN is able to share the radio resources for RABs, having different QoS. Therefore, more sophisticated mechanisms need to be utilised to exploit the radio resources efficiently and to meet the diverge QoS requirements of the RABs as well as is possible. To respond to this demand, the advanced RRM algorithms must be accompanied with more adaptive MM as it was for 2G.

As a result, the concepts of RRC state transition and hierarchical MM have been specified for UMTS, including UTRAN level MM, which is new compared to the GSM MM, which was taken care of by the CN subsystem. Based on this approach, during the radio connection the UE can have different states depending on the type of connection it may have to the UTRAN as well as the motion speed of the UE. It is the responsibility of the RNC to control the UE states by considering the UE mobility, the requested RAB(s) and its variation in terms of bit rate.

The cornerstones of the UTRAN level mobility are based on the concept of cell, UTRAN Registration Area (URA), Radio Network Temporary Identifier (U-RNTI) and the RRC state-transition model. In addition, the primary purpose of defining the different logical roles for RNC and specifying Iur interface was also to support the UTRAN internal MM.

A *URA* is defined as an area covered by a number of cells. The URA is only internally known in the UTRAN and hence it is not visible to the CN. This means that whenever a UE has a RRC connection its location is known by the UTRAN at the accuracy level of one URA. Every time the UE enters a new URA it has to perform a URA updating procedure. Having the interface between RNCs facilitates the URA area that can cover in principle different RNC areas.

Based on the 3GPP specification TS 25.401 *Radio Network Temporary Identities* (RNTI) are used as UE identifiers within UTRAN and in signalling messages between UE and UTRAN. There are four types of RNTI used to handle the UTRAN internal mobility, including:

- SRNC RNTI (S-RNTI)
- Drift RNC RNTI (D-RNTI)
- Cell RNTI (C-RNTI) and
- UTRAN RNTI (U-RNTI).

S-RNTI is allocated in association with RRC connection set-up by the SRNC to which the UE has the RRC connection. By this means the UE can identify itself to the SRNC and also the SRNC can reach the UE.

D-RNTI is allocated by a DRNC in association with context establishment and is used to handle the UE connection and context over the Iur interface.

C-RNTI, which is allocated when the UE accesses a new cell, it is a CRNC specific identifier and used to identify the CRNC to which the UE has the RRC connection. By this means the UE can identify itself to the CRNC and also the CRNC can reach the UE. This identifier is used when the UE is in Cell Forward Access Channel (FACH) state.

The U-RNTI is, on the other hand, allocated to a UE having a RRC connection and identifies the UE within UTRAN. U-RNTI is used as a UE identifier for the first cell access (at cell change) when a RRC connection exists for this UE and for UTRAN originated paging including associated response messages.

Allocating and handling of the above-mentioned identifiers is a primary responsibility of RNC.

Having defined the RNC roles, UE identifiers, and the concept of URA, then the question is how to combine this with the state transition in order to handle the UTRAN internal mobility and RRM. The main principles of the way of handling the RRC state transitions are shown in Figure 4.31 and can be summarised as follows:

- *No radio connection:* the UE location is known only by the CN (in accuracy of Location and/or Routing Area, which are CN level location area concepts and are described in Chapter 5). This means that the location information is stored in the network based on the latest MM activity the UE has performed with the CN.
- *Radio connection over common channels:* if the radio connection uses common channels, for example, FACH, and CPCH, the location of the UE is known in accuracy of a cell. The way in which this location information is updated is RRC procedure Cell Update. This state can be used when there is low bit rate data to be transferred between the UTRAN and the UE.
- *Radio connection over DCHs:* in this case the UTRAN has allocated dedicated resources for the connection (DPDCH) and Dedicated Physical Control Channel (DPCCH) as mini-

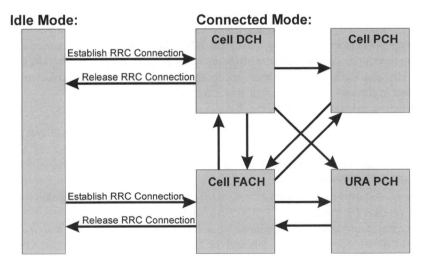

Figure 4.31.   RRC state automaton

mum, number of DPDCH depends on the used bandwidth). The location of the UE is known on a cell level. Depending on the type of connection, different RRC procedures may occur. If the dedicated connection is used to carry a service using the highest QoS class (for example, circuit switched voice call is one of these) the UTRAN and the UE perform handovers. If the service the radio connection carries uses lower QoS class, for example, web surfing, allowing buffering and delays, handovers may not be completed as such. Instead, the UE uses Cell Update procedure to inform the UTRAN about its location, i.e. the place where the data should be delivered over the radio connection.

- *Radio connection in Cell PCH state:* Once the UE is in the Cell FACH or Cell DCH state but there is no data to be transferred, the UE state is changed to the Cell PCH state at which it can monitor the paging occasions based on the defined Discontinuous Reception (DRX) cycles and hence hear the paging channel. In this state the location of the UE is known at the accuracy of a single cell sometimes called the home cell.
- *Radio connection in URA PCH state;* Once the UE is in the Cell FACH or Cell DCH state and there is no considerable data to be transferred between the UE and the UTRAN, or the UE mobility is high, then its state can be transferred to the URA PCH state in order to avoid periodical cell update and to release the dedicated radio resources, respectively. In this case the location of the UE is known only at an URA level and hence in order to obtain the cell level location accuracy the UE should be paged by the UTRAN/RNC. This is the case, for example in association with mobile positioning process when cell based positioning method is used. In this state the UTRAN benefits from the advantage of having Iur interface between the RNCs and the URA concept.
- *Idle mode;* The RRC Idle Mode equals the state where the UE and the UTRAN do not have any radio connection, for example, the UE is switched off. In this RRC state the network does not have any kind of valid information concerning the location of the UE.

4.3.2.6 Database Handling

Like GSM BSC, the RNC contains stores of information that can be called a radio network database. This database is the data storage for cell information. The cell information for the cells RNC controls is stored in this database. The RNC then sends this cell-related information to the correct cell, which further distributes it by broadcasting over the Uu interface towards the UE. The radio network database contains plenty of cell-related information and the codes used in the cell are only part of this information.

The information in a radio network database can roughly be classified as:

- Cell Identification information: Codes, Cell ID number, Location Area ID and Routing Area ID of the cell.
- Power Control information: allowed power levels in the uplink and the downlink directions within the cell coverage area.
- Handover related information: connection quality and traffic related parameters triggering handover process for the UE.
- Environmental information: the neighbouring cell information (both GSM and WCDMA). These cell lists are delivered to the UE and the UE performs radio environment measurements as a preliminary works for the handovers.

4.3.2.7 UE Positioning

Another important function the RNC handles is to control the UE positioning mechanism in UTRAN. In this association it selects the appropriate positioning method and controls how the positioning method is carried out within UTRAN and in the UE. It also co-ordinates the UTRAN resources involved in the positioning of the UE.

In network-based positioning methods the RNC calculates the location estimate and indicates the achieved accuracy. It also controls a number of Location – Measurement Units/BSs (LMUs/BSs) for the purpose of obtaining radio measurements to position or help to position UE.

Since the UE positioning is seen as a value-added service in UMTS networks we discuss it more thoroughly in Chapter 7 in conjunction with the UMTS Services.

5

UMTS Core Network

The UMTS Core Network (CN) can be seen as the basic platform for all communication services provided to the UMTS subscribers. The basic communication services include switching of circuit-switched calls and routing of packet data. Value-added services on top of these basic services are discussed in Chapter 7.

The CN maps the end-to-end Quality of Service (QoS) requirements to the UMTS bearer service. When inter-connecting to the other networks the QoS requirements also need to be mapped onto the available external bearer service. This gateway role of the UMTS CN in creation of the end-to-end service path is illustrated in Figure 5.1. The external bearer is not in the scope of the UMTS system specifications and this may create some local problems if the QoS requirements to be satisfied between the UMTS and external network do not match.

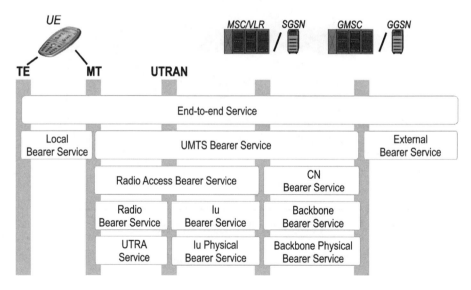

Figure 5.1. Bearer and QoS architecture in CN

Between the MT and CN the QoS is provided by the radio access bearer. The radio access bearer hides the QoS handling over the radio path from the CN. Within the CN the QoS requirements are mapped to its own bearer service, which in turn is carried on the backbone bearers on top of the underlying physical bearer service. A challenge in the CN implementation is that the operator has pretty much freedom in choosing how to implement the physical backbone bearers. The physical backbone bearers rely on the physical transmission technologies used between the CN nodes. Typical transmission technologies like PDH and SDH with PCM channelling or with ATM cell-switching are selected. These technologies and protocols are briefly introduced in Chapter 9.

UMTS represents a kind of philosophy to produce a universal core, which is able to handle a wide set of different radio accesses. Referring to the network evolution discussed in Chapter 2, the new wideband radio access is one of the evolution steps, which upgrades the whole network with new facilities and capabilities. The core part of the UMTS network does not evolve in as straightforward a way as the radio network due to both the traditional infrastructure basis and those advanced technologies which may have various impacts on the evolution of the core part of UMTS. Figure 5.2 shows the conceptual nature of the UMTS CN: the radio accesses drawn with a continuous line are the ones used in the beginning and the others are considered as access candidates as time goes by.

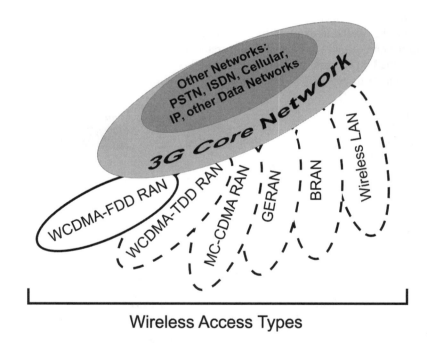

Figure 5.2.   Universal core for wireless access

The development vision presented in Figure 5.2 takes its first steps in 3GPP R99; the palette of radio accesses is expanded in 3GPP R4 and R5 and further steps are under consideration. Nevertheless, general advancements in networking technologies are impacting on the UMTS CN itself and the fairly simple CN architecture described here is already undergoing changes as will be shown at the end of this chapter.

5.1. CN Architecture in 3GPP R99

The first version of UMTS Specifications, 3GPP R99, introduces a system having a wideband radio access and a CN evolved from GSM. This means, that the original GSM platform with GPRS extensions for packet data services should be used as effectively as possible. This is an obvious approach since every party involved is aware of the costs the establishment of UMTS will generate. By utilising the existing platform(s) every party involved is in a position to save some of the costs. Since the network subsystem of the GSM/GPRS is capable of providing the basic communication services for both circuit- and packet switched traffic together with a rich set of supplementary and value-added services, it became an obvious choice for the basis of the UMTS CN.

Also in the 3GPP R99 CN the traffic will be either circuit switched or packet switched in nature. Both of these traffic types require some specific arrangements and this is why the CN is functionally further divided into two domains, circuit switched domain (CS domain) and packet switched domain (PS domain). The CN architecture is shown in Figure 5.3.

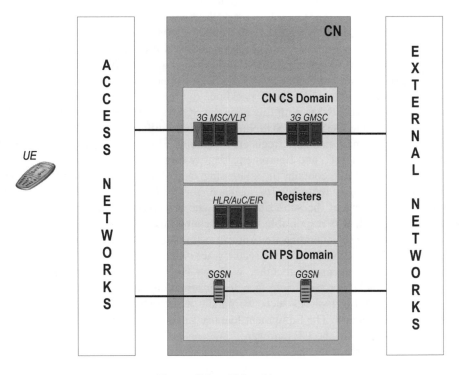

Figure 5.3.   CN architecture

The CN CS domain has two basic network elements, which can be physically combined. These elements are serving Mobile Switching Centre/Visitor Location Register (MSC/VLR) and Gateway Mobile Switching Centre (GMSC). The serving MSC/VLR element is responsible for circuit switched connection management activities, mobility management (MM) related issues like location update, location registration, paging and security activities. The

GMSC element takes care of the incoming/outgoing connections to/from other networks. From the connection management point of view, the GMSC establishes a call path towards the serving MSC/VLR under which the addressed subscriber is to be found. From the MM point of view, the GSMC initiates a location info retrieval procedure, whose aim is to find the correct serving MSC/VLR for call path connection.

In addition, the MSC/VLR entity contains the transcoders required for speech coding conversion. The remarkable difference here is that in the GSM specifications the transcoders were considered as a part of the radio network, in 3G they are part of the CN. This is an indication of the development, where delay sensitive processing, like speech coding, is continuously moving towards the border between the UMTS network and other networks.

The CN PS domain has also two basic mobile network specific elements, Serving GPRS Support Node (SGSN) and Gateway GPRS Support Node (GGSN). The SGSN node supports packet communication towards the access network. If GSM BSS is in question, the interface is called Gb and if UTRAN is in question, the interface is Iu. The SGSN is mainly responsible for MM related issues like routing area update, location registration, packet paging and controlling the security mechanisms related to the packet communication. The GGSN node maintains the connections towards other packet switched networks such as the Internet. From the CN point of view, this node is responsible for MM related issues like GMSC in the CN CS domain. The session management responsibility is also located in the GGSN.

The transport network connecting GSNs together is called *IP backbone,* which can be regarded as a private "Intranet". This is why the IP backbone is actually separated from other networks by firewall functionality. For IP backbone routing the PS domain must also contain a Domain Name Server (DNS). With this node the SGSNs and GGSNs are able to perform routing and actually the GGSN and SGSN may belong to different networks in this respect.

In Figure 5.3 the part named "Registers" contains Home Location Register (HLR), Authentication Centre (AuC) and Equipment Identity Register (EIR). This part of CN does not deliver traffic. Instead it contains the addressing and identity information for both the CS and PS domain, which is required, for instance, for MM related procedures. The HLR contains permanent data of the subscribers. One subscriber can always be registered into only one HLR. The HLR is responsible for MM related procedures.

The AuC is a database generating the Authentication Vectors. These contain the security parameters the VLR and SGSN use for security activities performed over the Iu interface. Typically the AuC is integrated together with the HLR and they use the same Mobile Application Part (MAP) protocol interface for information transfer. Security is a part of MM but since it is a big topic as such, we handle it separately. For UMTS environment security, refer to Chapter 8. The Equipment Identity Register (EIR) maintains the identification information related to the UE hardware.

In addition to these registers, the CN contains one more register, Visitor Location Register (VLR). In GSM, this register was handled separately in the specifications but in 3G the VLR is considered to be an integral part of the serving MSC. The VLR participates in MM related procedures like location update, location registration, paging and security activities. The VLR database contains temporary copies of the active subscribers, which have performed location update in the VLR area.

As explained earlier, the CN contains two separate domains for traffic delivery and these domains take into account the special characteristics of the traffic. The special characteristics of the traffic also effect the CS and PS domain element addressing scenarios and further on the signalling interfaces and their transport. The CS domain uses GSM inherited signalling scenarios based on a Mobile Application Part (MAP) protocol covering any possible add-ins the UMTS brings into the system. In Figure 5.4 these inherited interfaces are marked with capitals according to the MAP interface naming rules. These interfaces follow the same functioning principles as already used in the GSM system.

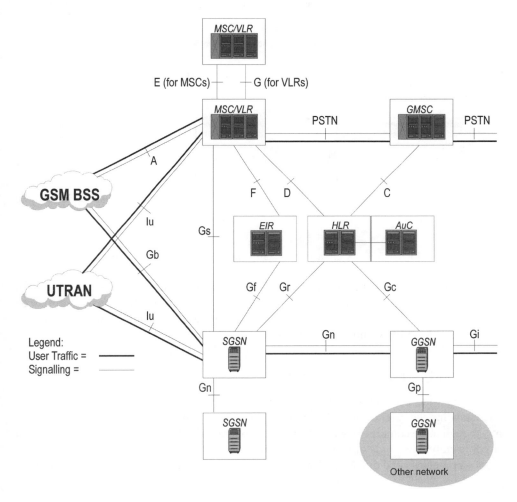

Figure 5.4.   3GPP R99 CN interfaces

The PS domain in UMTS is evolved from the 2G GPRS and this can be seen from the inherited interface names always starting with G and having another letter indicating which interface is in question. From the functional point of view the Gx interfaces resemble their CS domain counterparts. For instance, the interface Gc uses similar procedures and mostly the same parameters as the MAP interface C; both of these interfaces retrieve location informa-tion from the HLR.

The rest of this chapter represents the CN related management tasks and control duties as shown in Figure 5.5. As far as Communication Management (CM) is concerned, the two main tasks are connection management and session management. The connection management is a management task responsible for circuit switched transactions and related issues (refer to Section 5.3). The control protocols carrying CM information, which deal with call and session control, are referred to here as a set of Communication Control (COMC) protocols.

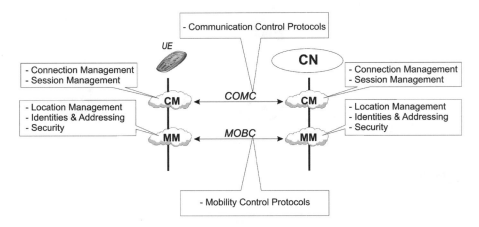

Figure 5.5.   Core network management tasks and control duties

The MM task covers management of UE locations together with their identities and addresses related issues and also security is considered as part of MM. Security is discussed in Chapter 8 in more detail. The control protocols supporting execution of MM tasks are referred to as Mobility Control (MOBC) protocols.

5.2. Mobility Management (MM)

With world-wide 2G cellular networks mobility is here to stay in communication networks. Understanding the essence of mobility makes the mobile network design significantly different – though more complex as well – from fixed communications and creates a lot of potential for completely new kinds of services to the end users.

Let's first clarify the difference between two basic concepts related to users' mobility:

- Location
- Position

The term "location" is used to refer to the location of the end user (and his/her terminal) within the logical structure of the network. The identifiable elements within such a logical structure are the cells and areas composed of groups of cells. Please note that the word "area" does not need to refer to a set of geographically neighbouring cells, but is composed by the network operator for network operation purposes.

The term "position" on the other hand refers to the geographical position of the end user (and his/her terminal) within the coverage area of the network. The geographical position is given as a pair of standardised co-ordinates. In the most elementary case, when no geographical position can be determined, the position may be given as a cell identity, from which the position can be derived, e.g. as the geographical co-ordinates of the BS site controlling that cell.

Although both location and position answer to the question, where is this user, the answers are used by the UMTS network in a completely different manner. The location information is used by the network itself to reach the users whenever there is communication service activity addressed to them. The position information is determined by the UMTS network when requested by some external service, e.g. Emergency Call Centre. Although the position information may well be life-critical to the end user making the emergency call, the location information is "life-critical" to the network itself in being able to provide services to the mobile users in a non-interrupted manner.

It should be noted that the primary purpose of positioning is to support application-oriented services, position information could also be utilised internally by the network. Examples of internal applications are position-aided handover and network planning optimisation.

Mobile positioning as a service capability is introduced among other services in Chapter 7.

Another key service created by the MM is *roaming*. The MM functions inside a single PLMN allowing a UMTS user to move freely within the coverage area of that single PLMN. Roaming is a capability, which makes it possible for the users to move also from one PLMN to another operated by a different operator company and possibly even in a different country. For the purpose of roaming many of the interfaces shown in Figure 5.5 (especially C, D, Gc and Gr) are inter-operator interfaces, which are used by the CN elements in visited serving networks to retrieve subscriber location and subscription information from their home networks.

As stated, MM needs a kind of logical, relative hierarchy for its functioning. In addition to this structure, the MM handles identities (permanent and temporary) and addressing information of the subscribers and their terminals as well as those of the involved network elements. In the specifications the security aspects are also counted as part of mobility management.

5.2.1. Identities and Addressing of Users and their Terminals

Unlike in fixed networks, the UMTS network requires many kinds of numbers and identities to be used for different purposes. In fixed networks the location of the subscriber and the equipment is, like the name says, fixed and this in turn makes many issues constant. When the location of the subscriber is not fixed, the fixed manners of numbering are not valid any more. The purposes of different identities used in UMTS can be summarised as follows:

- Unique identity: this is used to provide a globally unique identity for a subscriber. This value acts as primary search key for all registers maintaining subscriber information and it is also used as a basis for charging purposes.
- Service separation: especially in the case of mobile terminating transactions the service going to be used must be recognised. This is done by using an identity having a relationship to the unique identity of the subscriber.
- Routing purposes: some special arrangements are required in order to perform transaction routing which is not fixed to any network and country borders.
- Security: security is a very important topic in the cellular environment and this is why additional identities are generated to improve the privacy of users. Basically, these security related identities are optional but it is strongly recommended to use them anyway.

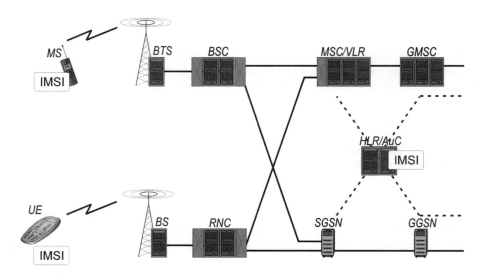

Figure 5.6.   International mobile subscriber identity (IMSI)

The unique identity for the mobile subscriber is called *International Mobile Subscriber Identity* (IMSI) (Figure 5.6). IMSI consists of three parts:

$$IMSI = MCC + MNC + MSN$$

Here MCC is the mobile country code (three digits), MNC the mobile network code (2–3 digits) and MSN the mobile subscriber number (9–10 digits). This number is stored in the SIM card (USIM). The IMSI acts as a unique database search key in the HLR, VLR, AuC and SGSN. This number follows the ITU-T specification E.214 on numbering. When the mobile user is roaming outside the home network, the visited serving network is able to recognise the home network by requesting the UE to provide this number. Because the IMSI is a unique database key for the subscriber's HLR located in the subscriber's home network, the HLR is able to return the subscriber profile and other information when requested by IMSI. The same procedure is applied for security information requests from the home network.

The IMSI number is used for the exact subscriber identification. The *Mobile Subscriber ISDN Number* (MSISDN) is then used for service separation. Because one subscriber may have several services provisioned and activated, this number acts as a separator between these. For instance, the mobile user may have one MSISDN number for speech service, another MSISDN number for facsimile and so on. In the case of mobile originated transactions the MSISDN is not required for service separation because the indication of the service is provided within the CM message(s) during the transaction establishment. In the mobile terminated direction different MSISDN numbers are required for different services because the surrounding networks are not necessarily able to provide the service information by other means. The MSISDN consists of three parts:

$$MSISDN = CC + NDC + SN$$

where CC is the country code (1–3 digits), NDC the national destination code (1–3 digits) and SN the subscriber number. This number format follows the ITU-T specification E.164 on numbering. Very often this number is called "directory number" or just simply "subscriber number".

The functional packet switched counterpart for MSISDN is *Packet Data Protocol (PDP) context address*, which is an IP address of the mobile user. The PDP context address can be either dynamical or static. If it is dynamic, it is created when a packet session is created. If it is static, it has been defined in the HLR. If the PDP context address is static it behaves like MSISDN on the circuit switched side of the network (see Figure 5.7).

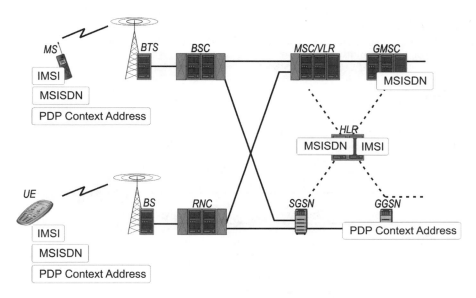

Figure 5.7.   Mobile Subscriber ISDN Number (MSISDN)

The *Mobile Subscriber Roaming Number* (MSRN) is used for call routing purposes (Figure 5.8). The format of the MSRN is the same as MSISDN, i.e. it consists of three parts, CC, NDC, SN and it follows the E.164 numbering specification.

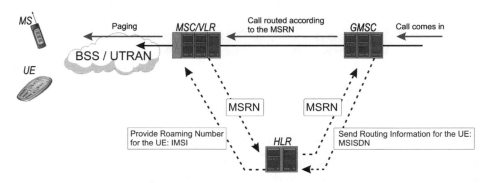

Figure 5.8.   Mobile Subscriber Roaming Number (MSRN)

MSRN is used in mobile terminated call path connection between the gateway MSC and serving MSC/VLR. This is possible since the allocated MSRN number recognises country, network and also the network element within the network. The "subscriber part" of MSRN is for subscriber recognition. The MSRN is also used for call path connection between two MSC/VLRs in the case of MSC-MSC handover. In this context the MSRN is often called *Handover Number* (HON).

Due to security reasons it is very important that the unique identity IMSI is transferred to non-ciphered mode as seldom as possible. For this purpose, the UMTS system makes use of the *Temporary Mobile Subscriber Identity Number* (TMSI) instead of the original IMSI (Figure 5.9). The PS domain of the CN allocates similar temporary identities for the same purpose. In order to separate those from the TMSI, they are called *Packet Temporary Mobile Subscriber Identities* (P-TMSIs).

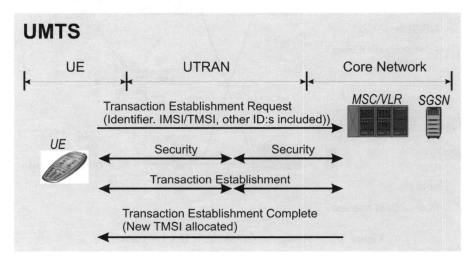

Figure 5.9.   Allocation of Temporary Mobile Subscriber Identity (TMSI)

The TMSI and P-TMSI are random-format numbers, which have limited validity time and validity area. The TMSI numbers are allocated by the VLR and they are valid until the UE performs the next transaction. The P-TMSI is allocated by the SGSN and it is valid over the SGSN area. The P-TMSI is changed when the UE performs routing area update.

International Mobile Equipment Identity (IMEI) is a number uniquely identifying the UE's hardware (Figure 5.10). There is a separate register called EIR (Equipment Identity Register) handling these identities. The UE provides the IMEI number within the transaction establishment request in addition to the other identities. The CN may or may not perform an IMEI checking procedure with EIR.

5.2.2. Location Structures and their Identities

In addition to the addressing and identities of the subscribers and their terminals the MM requires logical structuring of the network to be defined. This logical structure is represented as logical parts of the access network. Thus, these logical entities act like a "map" for MM

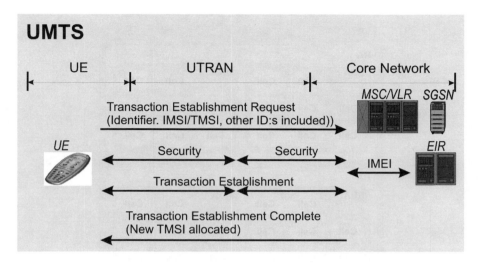

Figure 5.10.   International Mobile Equipment Identity (IMEI)

procedures and their parameterisation. UMTS contains basically four-level logical definitions, which can be listed as:

- Location Area (LA)
- Routing Area (RA)
- UTRAN Registration Area (URA)
- Cell

In the CN CS domain the LA is the area where the UE may move without performing the
location update procedure. The LA consists of cells; the minimum is one cell and the
maximum is all the cells under one VLR. In the location update procedure the location of
the UE is updated in the VLR with LA accuracy. This information is needed in the case of
mobile terminated calls; the VLR pages the desired UE from the location area in which it has
performed the latest location update.

It should be noted that in other respects other than the VLR the LA does not have any other
hardware bindings. For instance, one RNC may have several location areas or one location
area may cover several RNCs. Every location area is uniquely identified with *Location Area
Identity* (LAI). The LAI consists of the following parts:

$$LAI = MCC + MNC + LA \text{ code}$$

where the MCC and MNC are of the same format as in the IMSI number. The LA code is just
a number identifying LA. LAI is a unique number throughout the world and within the same
network the same LA code should not be repeated. One VLR cannot handle duplicate LA
codes. The UE listens to the LAI(s) from the transport channel BCH. The content of this
transport channel is cell-specific and is filled by the RNC.

Like the CN CS domain, the PS domain has its own location registration based on RA. RA
is very similar to LA, i.e. it is the area where the UE may move without performing the RA
update. On the other hand, the RA is a kind of "subset" of LA: one LA may have several RAs
within it but not vice versa. In addition, one RA cannot belong to two location areas.

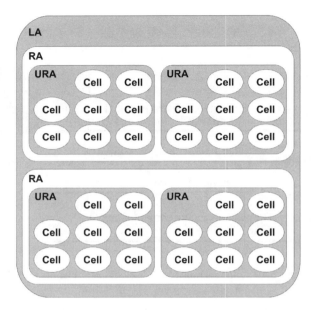

Legend:
LA   = Location Area
RA   = Routing Area
URA = UTRAN Registration Area

Figure 5.11.   MM logical entities and their relationships

The VLR and SGSN have an optional interface Gs between them (refer to Figure 5.4) through which these nodes may exchange location information. Since UMTS must interoperate with GSM, the UMTS CN supports also features available in GSM. One of these features is a combined location/RA update, where the GSM terminal performs updated requests first to the SGSN. If optional Gs interface is available, the SGSN also requests through this interface the VLR to update location area registration. In the plain UMTS network the combined location/RA area update is not available and a UE registers its location to both CN domains separately.

In the GSM network the MM is completely handled between the terminal and NSS. In UMTS the UTRAN is partially involved in MM. Due to this the UTRAN contains local mobility registration, a *UTRAN Registration Area*(URA) as discussed in Chapter 4. This "sounds-like-a-small" change causes remarkable changes in the SGSN internal structure and actually, because of this, the 3G SGSN contains both 2G SGSN and 3G SGSN functionality. In UMTS the SGSN carries tunnelled IP traffic to/from a UE according to the URA identity. In 2G the SGSN terminates the tunnelled IP traffic and relays it over the 2G-specific Gb link.

Because URA is almost a similar kind of logical definition than LA and RA, it does not have, in principle, any limitations in respect of network elements. In practice it seems to be so that URA and radio network subsystems (RNS) have more or less fixed relationships. On the other hand, URA is, in a way, a logical definition, which combines traffic routing and radio resource control. In routing the URA the addressing entity is pointing towards the access domain and in radio resource control the terminal has states indicating the location accuracy and traffic reception ability. This is visible in the RRC state model, which was briefly discussed in Chapter 4.

The smallest "building block" used for the preceding MM logical entities is a cell. Basically the CN does not need to be aware of cells directly, but sets of cells, i.e. areas. The cell in the access domain is the smallest entity having its own publicly visible identity called *Cell ID* (CI). Like the LA code the CI is also just a number, which should be unique within the network. To globally separate cells from each other, the identity must be expanded and in this case it is called *Cell Global Identity* (CGI). The CGI has the following format:

$$CGI = MNC + MCC + LA \text{ code} + CI$$

The CGI value covers the country of the network (MCC), the network within a country (MNC), location area in the network and finally the cell number within the network. This information is distributed to the UE by the UTRAN functionality for system information broadcasting.

5.2.3. Mobility Management State Model

The presence of packet connection and its management brings in a new dimension as far as MM is concerned; for packet connections the MM has a state model. In circuit switched connections basically the same kind of model exists but it is rarely used because the circuit switched connection behaviour does not require this kind of state model. As already indicated in Figure 5.5, the abbreviation MM refers to the circuit switched MM. The abbreviation GMM refers to packet switched MM.

5.2.3.1. MM States in Circuit Switched Mode

From the MM point of view a terminal, with respect to network connection, may have three states, MM-detached, MM-idle and MM-connected. These MM states indicate how accurately the terminal location is known when comparing to the logical structure presented in Figure 5.11. In the MM-detached state the network is not aware of the terminal/subscriber at all. This is the MM state, for instance, when the terminal is switched off. In the MM-idle state the network knows the location of the terminal/subscriber with the accuracy of LA. In the MM-connected state the network knows the location of the terminal with the accuracy of a cell.

The situation described in Figure 5.12 is similar both in the GSM NSS and in the UMTS CS domain following the 3GPP R99. When a subscriber switches his/her terminal on, the terminal performs either IMSI attach or location update procedure. Accordingly, when the subscriber switches his/her terminal off the MM state changes from MM-idle to MM-detached. IMSI attach is performed if the LAI recognised from the camped cell is the same as the terminal has in store in the USIM. If the LAI received from the network differs from the stored one, the terminal performs location update in order to update and possibly register its new location within the CN CS domain and HLR. In either case the MM state changes as follows: MM-detached → MM-connected → MM-idle. The reason here is that when the terminal performs either of these shortly described procedures the network is momentarily aware of the terminal location with the accuracy of a cell. Both of these

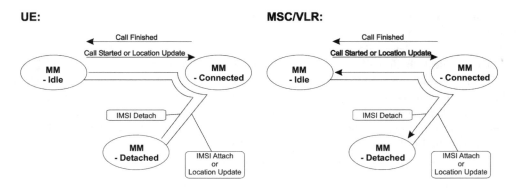

Figure 5.12.   Circuit switched MM state model (MM)

procedures "wake up" the circuit switched domain with same type of message containing the originating cell information and reason for the transaction. In GSM this message is called "CM Service Request" and in UMTS it is "UE Initial Message".

When a subscriber is active (MM-idle, terminal switched on) the MM state is toggled between MM-idle and MM-connected states according to the use of the terminal. To put it simply: when a call starts the MM state goes MM-idle → MM-connected and when the call is finished the MM state goes MM-connected → MM-idle.

5.2.3.2. MM States in Packet Switched Mode

For the packet switched connections the situation is different in terms of PMM procedures: The PMM (Packet Mobility Management) states are the same but the triggers that allow movement from one state to another are different (Figure 5.13).

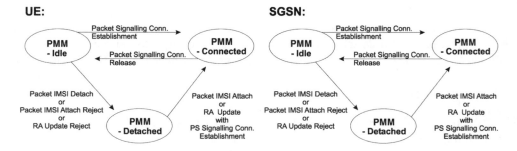

Figure 5.13.   Packet switched MM state model (PMM)

When the PMM state is PMM-detached the network does not have any valid routing information available for the packet switched connection. If the PMM state should be changed an option is to perform packet IMSI attach. This procedure takes place whenever a terminal supporting packet switched operation mode is switched on. A confusing issue here is that this so-called packet attach is a very different procedure than its counterpart on the circuit switched side of the network. Packet IMSI attach as a procedure is a relatively heavy signalling procedure reminding location update. With packet IMSI attach the valid routing information for the packet switched connection is "created" in every node involving the packet switched connections, SGSN and GGSN. In addition to these, the subscriber profile is requested from the HLR and possible old routing information is eliminated.

In the PMM-connected state the data can be transferred between the terminal and the network: the SGSN knows the valid routing information for packet transfer with the accuracy of the routing address of the actual serving RNC. In the PMM-idle state the location is known with the accuracy of a routing area identity. In the PMM-idle state the paging procedure is needed in order to reach the terminal, e.g. for signalling.

From the end user point of view, the packet switched mobile connection is often described as "being always on" and on the other hand, a packet call is said to be like many short circuit switched calls. Both of these statements contain some truth but they are not quite exact as such. From the network point of view, the packet switched mobile connection gives an illusion of "being always on". This illusion is created with the MM states PMM-connected and PMM-idle. In the PMM-idle state both the network and the terminal hold valid routing information and they are ready for packet data transfer but they are *not* able to transfer packets in this state since there isn't any connection present through the access network.

When the subscriber switches his/her packet-transfer-capable terminal off, the MM returns back to the state PMM-detached and the routing information possibly present in the network nodes is not valid any more. If, for one reason or another, errors occur in the context of packet IMSI attach or RA update, the MM state may return to PMM-detached.

5.3. Communication Management (CM)

In this subsection we briefly describe the main functions of CM in terms of both circuit switched and packet switched communications. From a circuit switched standpoint, the functionality is referred to as CM. From a packet switched standpoint, the entire functionality is referred to as session management. These functionalities are covered by presenting the main phases of a connection or session management process as well as the entities.

5.3.1. Connection Management for Circuit Switching

Connection management is a high-level name describing the functions required for incoming and outgoing transaction handling within a switch. Generally speaking, the switch should perform three activities before a circuit switched transaction can be connected. Those activities are number analysis, routing and charging. Connection management can functionally be divided into three phases, which the transaction attempt must pass in order to perform through-connection (Figure 5.14).

Number analysis is a collection of rules on how the incoming transaction should be handled. The number of the subscriber who initiated the transaction is called the calling number and the number to which the transaction should be connected is the called number. Number analysis investigates both of these numbers and makes decisions based on the rules defined. Number analysis is performed both in connection management Phase I and Phase II. In Phase I the switch checks whether the called number is reasonable at all and if any restriction such as call barring is to be applied with to calling number.

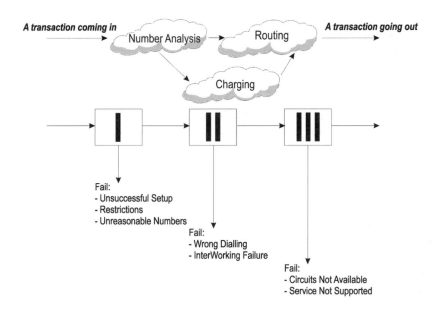

Figure 5.14.   Circuit switched connection management – principle diagram

In Phase II the system concentrates on the called number. The nature of the transaction is investigated: is it an international or national call and is there any routing rule defined for the called number at all? In addition, the system checks if the transaction requires any inter-working equipment (like a modem) to be connected and whether the transaction is chargeable or not. Also statistics for this transaction are initiated in this phase.

As a successful result of connection management Phase II, the system knows where the transaction attempt should be connected. This connection and channel selection procedure is called routing. When the correct destination for the transaction is known the system starts to set up channel(s)/bandwidth towards the desired destination by using, for instance, ISUP (ISDN User Part) signalling protocol. During the transaction the switch stores statistical information about the transaction and its connection and collects charging information, if the transaction was judged to be chargeable. When the transaction is finished, connection management Phase III takes care of releasing all the resources related to the transaction.

In fixed networks every call is treated as an entity, from end-to-end. In cellular networks a term "call" can be interpreted in many ways. Every "call" consists of call legs and each leg thus defines a part of a call.

From the connection management point of view, every call consists at least two legs. There are four legs available MOC (Mobile Originated Call), MTC (Mobile Terminated Call), POC (PSTN Originated Call) and PTC (PSTN Terminated Call). As Figure 5.15 indicates, the connection management is actually a distributed functionality and, depending on which element is in question, different parts of the call control are used. If a Serving MSC/VLR is in question, it handles MOC and MTC call legs and if the GMSC is in question it handles POC and PTC legs. The call control is able to receive and create these call legs and also to determine whether any additional functionality is required based on the leg and call type. The most important additional facilities are network interworking and charging.

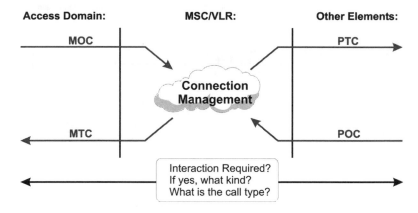

Figure 5.15.   Call legs

The call control recognises the call type and based on this it decides the further actions. The basic circuit switched call types are:

- Normal call (voice)
- Emergency call
- Data call (facsimile included)

In the MOC leg, this call type is included in the "CM Service Request Message" (GSM) and "UE Initial Message" (UMTS with UTRAN). In the POC leg, the call type is "hidden" in the Called Party Address (B Number); as it was explained in the context of MM, the service to be used is recognised in the terminating direction by the MSISDN number. Both in MTC and PTC legs the connection management determines whether any kind of interaction is required between the call legs.

Connection management contains numerous tasks but two of them are the most visible ones for the end users and those are briefly presented in the following subsections.

5.3.1.1. Connection management – Network Interworking

The UMTS network must be adapted to the existing, surrounding networks like PSTN and ISDN. In principle, the circuit switched data calls are the ones requiring adaptation but in practice every call going beyond the cellular network border requires interworking facilities to be used.

Nowadays and also in the near future the "killer application" in cellular networks is still voice, though its share of the traffic is expected to decrease. Voice as such is a very demanding service and its proper handling requires some special activities, one of those being echo cancelling. Echo cancelling is especially required for the calls originating from the cellular network and terminating to the PSTN. Figure 5.16 shows what is going on with respect to PSTN.

Every circuit switched call established either through BSS or UTRAN is digital and thus four-wire connection in nature; this means that the transmission directions are completely separated from each other. When interconnecting with ISDN the connection is also four-wire

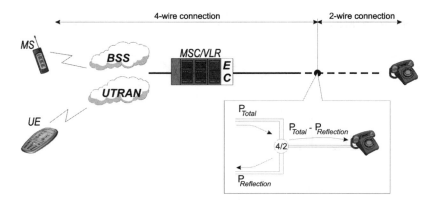

Figure 5.16.   Reflected power and where it comes from

connection on the ISDN side. The same applies to PSTN but there is one exception. In the PSTN networks the four-wire connection is converted to a two-wire connection (analogue one) in some point of the network. Typically this four/two-conversion point is very close to the PSTN subscriber. In other words, the so-called "last mile" (access network connection of the PSTN subscriber) is implemented as an analogue connection.

The coded voice samples sent from the cellular network are represented as power (P_{Total}). Because the four/two-conversion point is not ideal, a part of the power reflects back ($P_{\text{Reflected}}$) and the power going to the PSTN subscriber is the difference between these two powers. Depending on the PSTN connection quality and length, the reflected power arrives back to the cellular network after a certain time, which is expressed as delay. The reflected power can be heard as echo if it is delivered back to the calling cellular subscriber. To prevent this, the network uses Echo Cancelling equipment (EC) (Figure 5.17). The EC functioning principle is relatively simple: it stores the sent bit pattern and expects a similar bit pattern to be received within a certain, adjustable delay. When it arrives, the EC adjusts the signal phases and sums bit patterns together. If the phase adjustment is performed correctly, the result of this calculation in power is "0" and no "echo signal" is sent to the calling subscriber.

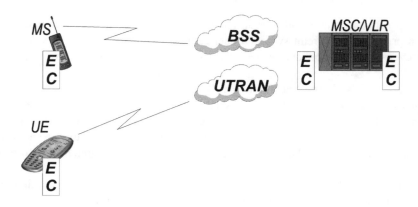

Figure 5.17.   Possible places for echo cancelling equipment

The above-described procedure is the way to eliminate echo in the case of network inter-working. This is not, however, the only source for echo. The other remarkable echo source is acoustic echo generated on the terminal and its operating environment. Acoustic echo cannot be handled with the same kind of system and this leads to the question what is the correct place for the EC in general?

In principle, the EC could be located in three places, the terminal, the edge of cellular access network or in the place described in Figure 5.16. From these alternatives, the best place is the terminal but this kind of implementation will make the terminal more complicated and thus more expensive. If the EC was located in the terminal the echo generated in the four/two-conversion point travels all the way back through the network and the echo is cancelled within the terminal. This makes the network side implementation easier but the fact is that the network still contains older cellular terminals unable to cancel echo. For those subscribers the network must offer echo cancelling on the network side. In other words, the echo cancelling functionality must reside in the network to guarantee equal service for all of the subscribers having any kind of terminal using voice service.

If a mobile-to-mobile call is performed no echo cancellers are normally involved since the connection is an end-to-end four-wire digital connection. In this kind of connection the acoustic echo is the problem. Acoustic echo can also be cancelled but the procedure is more complicated. The echo cancelling contains an algorithm with which the actual voice signal is separated from background noise and the background noise level is flittered as much as possible without causing any harm to the original signal. If this is desired, the echo cancelling equipment has again two alternative places; the EC functionality is either located in the terminal(s) or in the edge of the access network just before the MSC/VLR. EC will not be in the terminals because of the reasons stated above and thus the acoustic echo cancelling is done with the EC equipment located between the access network and the MSC/VLR. When done so, every voice call performed within the network uses echo cancelling for different purposes; in mobile-to-mobile calls for acoustic echo cancelling and in mobile-to-PSTN calls for delay based echo cancelling. From an echo handling point of view, the addition of echo cancellers between the access network and the MSC/VLR is an ideal solution but it contains some risk. Every time echo is cancelled it results in signal processing and processing always equals certain delay. Real-time types of services like voice are very sensitive for delays and thus any additional increasing of delay also increases risks from a connection quality point of view.

The other items the network interworking contains as its functionality are modems and rate adapters. Since the aim in UMTS is to transfer all kinds of data connections as packet switched, these parts of network interworking are quite marginal ones. The only exception here is the modems used for facsimile connections. The PSTN contains numerous amount of facsimile equipment using a circuit switched connection mode. To arrange traffic towards this equipment the UMTS network CS domain must contain adaptation facilities for this purpose.

In 3GPP R99 implementation the network interworking functionality is inherited from the GSM NSS and works, in principle, as such. In 3GPP R4 the same functionality is and must be present in the network but its implementation differs remarkably from the one used in 3GPP R99 implementation.

5.3.1.2. Connection management – charging

The existence of the non-real-time packet type of traffic generates remarkable change requirements as far as charging and accounting (billing) methods are concerned. Roughly, the UMTS network must be able to produce three different types of charging information acting as the basis for the billing. These three methods are time based charging, volume based charging and QoS based charging.

The UMTS system as such generates some new requirements concerning charging. Those changes can be short-listed as follows:

- Due to the changes in the commercial model (please refer to Chapter 7), the charging system must be able to produce detailed information in every leg of the commercial model, i.e. the charging happens between the subscriber and the network, the network and service provider and even between the service provider and content provider. It should be noted that these listed legs are not necessarily using the same network.
- Fraud control mechanisms are added/improved between the serving network and home network.
- The charged party may control the costs. Note: this has already been implemented in advanced 2G networks but in UMTS networks this will be a mandatory activity.
- Itemised billing must be possible for every subscriber and in every event. Note: an open question here is whether to charge or not a pure signalling transaction like location update.

Every transaction made in or through the UMTS network produces a detailed list of transaction related resources and this is called Call Detail Record (CDR). Normally one transaction, for instance, a voice call, produces several of these in every CN and intermediate network node involving the transaction establishment. The following list aims to describe what kind of items a CDR may include. Depending on the transaction type some of these items may be present or absent.

- *Charged party identity:* this is the IMSI number, which uniquely identifies the subscriber. This number also uniquely defines so-called home environment: IMSI contains mobile country and network portions.
- *Terminal identity and class:* the terminal identity is IMEI and class is described in the mobile station classmark. The mobile station classmark identifies the capabilities the UE is able to perform.
- *Called address:* this is also called B number and this identifies the destination and in the case of PSTN/ISDN also the service used over the UMTS network.
- *Used 3G resources:* bearer and its characteristics like bandwidth, identity and service type (connection oriented/connectionless).
- *QoS parameters:* especially in the CN PS domain the service pricing is based on the offered QoS. In practise, the QoS defines the delay type and value used in the connection.
- *Serving 3G network and network element identities:* for UMTS network, the MCC and MNC are unique identities and for the network elements the identities could be either SCCP level addresses or static IP addresses.
- *Time stamps:* when the connection has started and when it has been finished.
- *MM information:* this covers location and routing area identities and cell identities. These are especially used in the context of location based services.

- *CDR identity:* every CDR must be named so it can be uniquely identified over at least a 3-month period.

Billing as a term covers the post-processing system producing the payable bills from the CDR information collected from the UMTS network elements. The billing system itself does not belong to the UMTS network, only the interface is defined, i.e. what kind of information the CDR should contain.

The billing system and its flexibility is one of the key issues when implementing the UMTS network. Based on the experiences in 2G networks it can be stated that a new service can be technically taken into use quite easily and the major problems are related to the charging/billing verification and functionality concerning the service. The major challenge in UMTS will most probably be how to add volume based charging (CN PS domain) to the existing billing systems which are able to support time based charging.

Accounting as a term means the procedures related to charging and billing information transfer between the networks. In 2G networks this is done many ways and these ways are more or less operator specific. Normally these methods are agreed at the same time when the roaming connection is established between the networks.

In UMTS, the roaming between the networks is expected to be automatised, in some cases even nationally (regulatory issue). This sets the requirement that the accounting procedures should be automatic, too. The ways to implement this kind of activity are mainly:

- *CDR automatic forwarding:* the charging information is transferred between the networks. Something like this is already in use in some countries between different networks. A good example might be Finland, where the charging information is transferred between the PSTN and cellular networks. In protocols, some of them contain mechanisms to transfer charging information. One of the protocols able to do this is ISUP.
- *Charging algorithm:* this algorithm contains information about the subscriber's charging and it is downloadable from the subscriber's home network. When the subscriber is active in a visiting network, this algorithm is downloaded from the home network and the charging is performed in control of this algorithm. This method is also called charging delegation.

Charging algorithm described briefly above is one of the reasons why CAMEL (Common Applications for Mobile networks – Enhanced Logics) is, in practice, a mandatory facility in UMTS networks. Roughly, CAMEL is an evolved version of IN and contains possibilities to transfer subscriber-specific services between networks. Other reasons why CAMEL is required in UMTS networks are mainly related to services, their portfolio and availability.

5.3.2. Session Management for Packet Communication

In the PS domain the packet connections are called sessions and they are established and managed by an entity called Session Management (SM).

SM as a logical entity has two main states, inactive and active. In the inactive state the packet data transfer is not possible at all and also the routing information (if it exists) is not valid. In the active state packet data transfer is possible and all valid routing information is present and defined.

The protocol used for packet data transfer during an active session is referred to with a generic name Packet Data Protocol (PDP). The design of the CN PS domain allows many different PDP protocols to be used. The most obvious case is to use the IP as a PDP, but other protocols like X.25 could also be supported.

The SM handles packet session attributes as contexts and the term used here is *PDP context*. The PDP context contains all parameters describing the packet data connection by means of end-point addresses and QoS. For example, a PDP context holds information such as allocated IP addresses, connection type and related network element addresses. When the session management is active, i.e. a PDP content exists, the user has also an IP address. From the service point of view, one PDP context is set up for one packet switched service with a certain QoS class. Thus, for instance, web surfing and streaming video over packet connection have their own PDP contexts.

UMTS as defined in 3GPP R99 uses the following QoS classes:

- Conversational class
- Streaming class
- Interactive class
- Background class

These classes are discussed in more detail with examples in Chapter 7.

The PDP context is defined in the UE and the GGSN and when it exists, it contains all relevant parameters defining characteristics for the packet connection as described earlier. The PDP context can be activated, deactivated or modified.

The activation of the PDP context causes the SM to change its state from inactive to active. This SM state change in turn means that the UE forming a packet session with the network holds valid, allocated address information and also the characteristics of the packet connection are defined (for example, QoS to be used). When the PDP context has been activated the UE and the network are able to establish a bearer for data transfer.

The deactivation of the PDP context causes the SM to change its state to inactive. When the SM state is inactive, the address information and packet session information the UE and the network may have is not valid any more. Thus, the UE and the network are not in the position to arrange any connections to transfer user data flows.

When the SM state is active and a PDP context exists, the PDP context can be modified. In this modification process the UE and the network renegotiate packet session characteristics. A typical topic for this renegotiation is the QoS class of the packet session.

The SM is a high-level entity and its activity depends on the lower level entities being PMM and RRC. If the RRC and PMM states are not suitable for active packet session, the PDP context is deactivated and the SM state is changed from active to inactive. This kind of situation occurs, for example, when the RRC changes its state from connected to idle. This state change triggers the PMM to change its state from connected to idle and referring to the SM state model in Figure 5.18, the PMM state change triggers the SM to change its state from active to inactive.

The meaning of the SM is to create an illusion of an "always on"-type of connection to the end user and this must be done with an effective way by saving network resources whenever possible; for instance, the RABs carrying user data are established when required and when there is nothing to transfer the RABs are cleared but still the packet signalling connection remains. Figure 5.19 shows how the different management and controlling entities within CN and UTRAN change their states during an example packet data flow. This example describes a situation, where the subscriber turns his/her terminal on and IMSI attach is performed. After

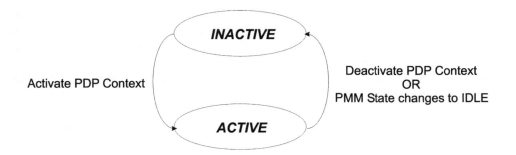

Figure 5.18.   SM state model

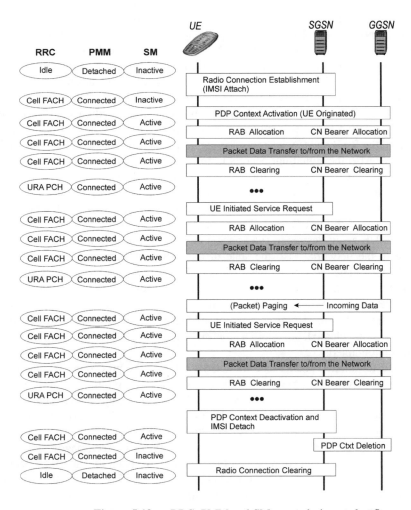

Figure 5.19.   RRC, PMM and SM states during packet flow

the IMSI attach the subscriber sends some packet data to the network and the service provider sends some data packets to the terminal. This procedure is repeated and then, after some time, the network sends some packet data to the UE. Finally the terminal is switched off. This example data flow can be realised, for example, during a WAP browsing session.

While the terminal is switched off there are no activities going on between the terminal and network. When it is switched on, an IMSI attach procedure is executed and the UE becomes identified by the network. The establishment of a signalling connection brings the RRC state from "idle" to "connected-cell FACH". At the same time the PMM state changes from "detached" to "connected" and the network has now valid information about the location of the subscriber. The SGSN gets this information in an "UE initial message" containing information about the desired activity. When the IMSI has been attached the UE initiates "PDP context activation". During this procedure the UE and the network negotiate desired packet connection characteristics, like for instance the QoS class. The result of the "PDP context activation" procedure is that the SM state changes from inactive to active.

If there is any packet data to be transferred a bearer able to carry user data flows is established. The SGSN starts the RAB allocation procedure over UTRAN and the CN bearer is established between the SGSN and the GGSN. Now the network is able to transfer packet data to and from the UE. The RRC state is used to optimise the UTRAN resources; when the RRC is in the "connectedcell FACH" state, only a small amount of packet data can be transferred over the Iu interface. In this RRC state the UE does not have a dedicated connection with the network. If the packet data amount was bigger, a dedicated channel would be allocated to the UE and the RRC state would be "connected-cell DCH". When the packet data has been transferred the RAB and CN bearers carrying user data flow are cleared but the PDP context is preserved. Also the signalling bearers forming the signalling connection between the UE and the network are maintained in this example. When the data bearers are cleared the RRC connection state changes to "connected- URA PCH" in order to save UE resources. In this RRC state the network does not know the exact location of the UE and if the network desires to communicate with the UE the UE must be paged.

After a while the UE again desires to send packet data to the network it sends a "service request" message to the network. Also the state of the RRC connection changes again to "connected-cell FACH". The "service request" triggers the network to allocate RAB and CN bearers. It should be noted that these bearers are allocated according to the parameterisation included in the PDP context. When these bearers have been established the packet data is transferred to and from the UE. Upon completion of the packet data transfer the bearers carrying user data flow are cleared, but the PDP context still remains active and the signalling connection is also maintained.

In the case of UE terminated packet data the GGSN initiates the procedure by sending a data packet to the SGSN currently serving the addressed UE. The reception of this data packet causes the SGSN to send packet paging to the desired UE. ‑Packet paging forces the UE to change the RRC state from "connected-URA PCH" to "connected-cell FACH" and the UE sends a "service request" message to the network. Since the SM is in the active state the network is able to allocate RAB and CN bearers according to the correct, negotiated parameters describing the packet connection. When the bearers have been allocated the packet data is transferred to the UE and, if the UE has packet data to be sent, it is sent to the network. When the packet data has been transferred the RAB and CN bearers carrying user data flow are cleared and only signalling bearers remain.

When the subscriber switches his/her terminal off, the "PDP context deactivation" takes place. This procedure removes any address information stored in the network concerning the packet connection and the PDP context is deleted. This causes the SM to change its state from active to inactive and packet data transfer is not possible any more. Since the UE is switching itself off, the signalling connection is not required any more and thus it is released. As a consequence, the PMM state goes to "detached" and RRC state is changed to "idle".

5.4. Architecture Aspects in 3GPP R4

In 3GPP R99 the major changes were targeted to the access part of the network; the main new issue presented was new wideband radio access, UTRAN. On the core network side the aim was to minimise changes and utilise the existing GSM/GPRS network elements and func- tionalities as much as possible. In 3GPP R4 the strategy is somewhat vice versa: the access network does not experience too much changes but the CN is extended remarkably.

Referring to Figure 2.9 in Chapter 2, especially the CN CS domain is to be changed. The leading principle in this respect is to separate CM and actual switching and related functions into separate physical entities.

The MSC/VLR evolves into MSC server and Media Gateway (MGW) (Figure 5.20). The MSC server is a network element containing CM main functionality, i.e. it maintains the logical CM structure presented in Figure 5.14. MSC server is also responsible for MM and MSC server also contains VLR. The MGW contains the facilities to perform actual switching and network interworking functions. For example, transcoders, echo cancelling equipment and modems are located at the MGW. Depending on the network configuration the MGW may contain plenty of other functionality, like for instance functionality to perform circuit packet conversions in the case of VoIP calls.

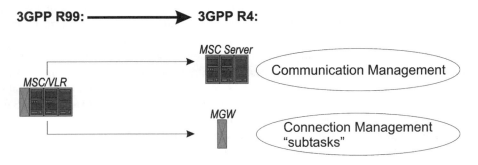

Figure 5.20.   MSC server and media gateway

The division between MSC server and MGW is not one-to-one; one MSC server may control numerous MGWs, this brings scalability into the system. On the other hand, one must take redundancy and security into account when planning the CN CS domain; if the MSC server amount is under dimensioned, the network may easily suffer from outages and rela- tively big amounts of subscribers will not gain circuit switched services.

From the CM point of view, the main issue is how to implement the connections between the UE and MSC server and between the MSC server and MGW and with which control protocol. There are several alternatives, but one of the most attractive one is IP based Session Initiation Protocol (SIP). SIP is also used in other connections with 3GPP R4/5. Current details of the SIP protocol can be found in the IETF RFC 2543.

As time goes by the IP is expected to be in use in as many instances as possible. Figure 5.21 presents a rough idea how IP conquers space from the other transport signalling, SS7. When the traffic transport is IP based, it is natural that traffic control should also be IP based, this is why SIP is seen as a very interesting alternative to implement call control protocols between the MSC server and MGW.

Figure 5.21.   The role of IP in 3G CN evolution

With user traffic and services the change is not necessarily so dramatic; there will be both packet and circuit switched traffic. The amount of circuit switched traffic is expected to decrease but the pace of this development is very hard to predict. The operators like to utilise the existing investments made to circuit switched technology first and when it is economically reasonable, the technology is changed.

From a transmission technology point of view, the circuit switched voice call wastes transmission resources. If we think of a normal voice call, its voice activity factor is, on average, 50%, i.e. one party is talking and the other one is listening. The system however reserves resources like the voice activity factor at 100%. Thus, transmission resources, on average, exceed the needs. Since transmission is a promising place to make cost savings, the Voice over IP (VoIP) has been a topic of increasing interest. By implementing VoIP and using IP as the transport technology, the transmission resources used for the call can be matched. Possible problems in this solution relate to QoS mechanisms. There are several ways to arrange QoS for real-time traffic but have not been tested in real life large-scale systems.

The VoIP can also be implemented in many ways. Since the UMTS network aims to be a global solution, the proprietary VoIP solutions are challenges in this respect. To tackle this issue, the 3GPP R4 contains an additional element called IP Multimedia Subsystem (IMS). The IMS, when fully specified, will contain a uniform way to maintain VoIP calls, thus offering to the operators a way to deliver VoIP calls between UMTS networks. This is not the only aspect of IMS. In addition to "universal VoIP" the IMS offers a platform to other real-time and non-real-time IP services, like for instance, multimedia services.

The IMS consists of Media Gateway Control Function (MGCF), Call State Control Function (CSCF) and MRF (Media Resource Function). These are functionalities and not network elements. These three functionalities form an extended model of CM when compared to 3GPP R99. Basically the MGCF controls the MGW used in the connections; conversions, echo cancelling, etc. The CSCF and MRF together form the logic of how the transaction using

IMS is treated. The logic with CSCF and MRF has relatively much analogy with the CM model presented in Figure 5.14.

The MGCF–CSCF–MRF model brings a big and new aspect to the system and this is related to the services, their creation and treatment. The 3GPP R4 contains service platforms (or service capabilities) and these are taken into use through the MGCF–CSCF–MRF chain. The basic concept exists in 3GPP R99 and actually the service capabilities are presented there, but in 3GPP R4 the CN structure supports the effective use of service capabilities. Service capabilities and related information are discussed more closely in Chapter 7.

In conclusion, the 3GPP R4 contains all the possibilities as far as traffic treatment is concerned. If the transaction coming from the access network is packet switched, it may be relayed to the external network either in packet switched or circuit switched manner. Also, if the transaction coming from the access network is circuit switched, it may go to the external network either in circuit or packet switched manner. In 3GPP R99 the nature of the connection (circuit/packet) remains the same throughout the network.

5.5. Architecture Aspects in 3GPP R5

In 3GPP R5 the access network will experience more changes and the changes in the CN are minor in nature. The main issues in 3GPP R5 are GSM/EDGE RAN (GERAN) and IP transport within the access network. In 3GPP R5 the traffic is always packet switched; here the question is whether it is real-time or non-real-time.

The reference architecture for 3GPP R4 and R5 is the same: refer to Figure 5.22. In the development of R5 the focus has shifted to the PS domain, which has been extended with the IMS functionality as described above. A more structured view on this part of the R5 system

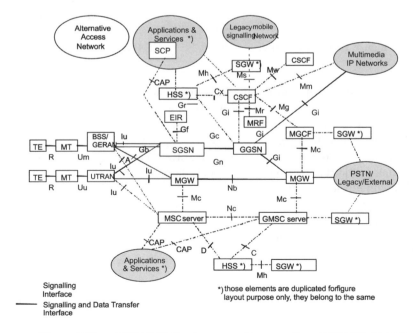

Figure 5.22.   3GPP R4/R5 reference architecture (from 3GPP TR 23.821)

with separate layers for user data transport, network control aspects and service capabilities is given in Figure 5.23. Referring to Figure 5.23, the services, their accessibility and creation are emphasised in 3GPP R4/R5 core network implementation. The service capability layer has already been introduced in 3GPP R99 but in further implementations its role will be increased by Open Service Architecture (OSA) based solutions. The OSA acts as a gateway to the fourth layer of the model presented in Figure 5.23 providing mechanisms for universal service creation and management. The service capability and application/service layer and their contents are briefly presented in Chapter 7.

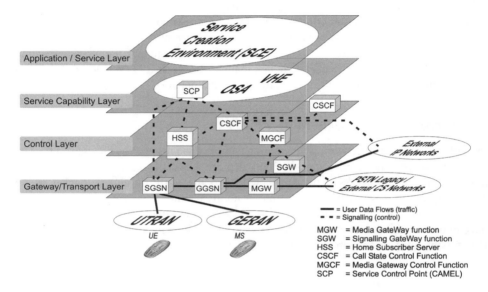

Figure 5.23.   Layered view of the 3GPP R5 CN (PS domain only)

The changes performed between R4 and R5 should not be visible to the end users. The UTRAN radio path still works like it has been working until now; also the terminals being used are still working as such. Within the access network the transport technology could be IP instead of ATM but this is the operator's choice.

Besides UTRAN the evolved GSM BSS named GERAN can be connected to the CN with Iu interface. Thus traffic coming from GERAN gets the same treatment as the traffic coming from UTRAN with regard to interfaces. If the operator has IMS in use, the CN CS domain is not basically needed any more; the main step (besides the new radio access alternatives) between 3GPP R4 and R5 is whether to quit the CN CS domain or not. This will depend significantly on the direction and maturing of the VoIP development.

6

The UMTS Terminal

The UMTS terminal is the most visible network element of the UMTS system as far as the end user is concerned. Therefore, it has been mostly considered as an element, which provides the application interface and services to the end user. Practically, however this is just the tip of the iceberg and the terminal function is considerably wider. The aim of this chapter is to provide a short insight into the UMTS terminal architecture and those function-alities beyond this architecture, bearing in mind the factors, which bring constraints and possibilities for the terminal architecture.

It should be noted that from network standpoint, that the overall protocol architecture and functions of the terminal are subject to standardisation. In spite of that their implementation and additional internal capabilities are strictly implementation dependent and hence seen as key competitive factors in the mobile communication business. Therefore, the architectural perspective presented in this chapter mainly rests on the specifications and is just one way to approach this interesting subject.

6.1. Terminal Architecture

The mobile end user's terminal side equipment of the radio interface is officially called a User Equipment (UE) in UMTS. The UE is often called according the GSM traditions a Mobile Station (MS) or shortly, a terminal. From the network point of view, the UE is responsible for those communication functions, which are needed on the other side of the radio interface, excluding any end user applications. The mandatory functionality of a UMTS terminal is related mainly to the interaction between the terminal and the network. The following func-tions are considered mandatory for all UMTS terminals:

- An interface to an integrated circuit card for insertion of the Universal Subscriber Identity Module (USIM)
- Service provider and network registration and deregistration
- Location update
- Originating and receiving of both connection-oriented and connectionless services
- An unalterable equipment identification (IMEI)
- Basic identification of the terminal capabilities
- The terminal must be able to support emergency calls without a USIM
- Support for the execution of algorithms required for authentication and encryption

Besides these mandatory functions, which are essential for network operation, the UMTS terminal should also support the following additional functionalities to facilitate future evolution:

- An Application Programming Interface (API) capability;
- A mechanism to download service related information (parameters, scripts or even software), new protocols, other functions and even new APIs into the terminal;
- Maintenance of the Virtual Home Environment (VHE) using the same user interface and/or another interfaces while roaming;
- Optional insertion of several IC cards.

The UE is often presented as one single monolith device mainly because the same vendor has delivered the physically indivisible equipment. But in sophisticated mobile systems the user equipment is often seen at least in standardisation as a set of interconnected modules with an independent group of functions. These modules may also be sometimes physically implemented as separate parts. In any case, these functional groups or subgroups have their own counterparts in the network side.

For example one of the main novelties in the GSM system is the separation physically of the user depending Subscriber Identity Module (SIM) and the general telecommunication system dependent part from each other by defining a standardised interface. This good idea is also inherited in UMTS. This idea enabled the subscriptions and operators as well as the physical terminal to be independent of each other.

Another important separation is the separation between the radio access network and core network dependent parts. However, this is not a matter of standardisation, but instead is due to the practical arrangements followed by suppliers when implementing the terminals. This modularity clarifies the manufacturing of multimode/multinetwork terminals at least internally and facilitates independence of the radio access technology from the core network technology. Figure 6.1 outlines the user equipment reference architecture as discussed here together with the related counterparts at the network side.

The UE consists of the ME, TE and the USIM.

The USIM is the user dependent part of the UE. The USIM is basically a logical concept and it is physically implemented into an integrated circuit card. The integrated circuit card, that may also contain application software or even multiple USIMs, is often called a Universal Integrated Circuit Card (UICC). The operator, with whom users make the subscription, will in any case provide the information content of USIMs. The USIM is connected rather to the service profiles than to a specific user profile. The ME is the user independent part of the UE, which consists of different modules.

The TE part of the UE is the equipment that provides the end-user application functions. It has to interact with the mobile termination part via the terminal adaptation (TA) function over the R reference point as shown in Figure 6.1. The terminal equipment terminates the telecommunication service platform. The TE knows the possible standardised telecommunication services on behalf of user applications. The terminal equipment may, for example, control the mobile termination by using the modem control command set (AT commands) defined by the ITU-T.

The Mobile Termination (MT) on the other hand is the part of the user equipment, which terminates the radio transmission to and from the network and adapts terminal equipment capabilities to those of the radio transmission. When looking from the mobile system point of view, the MT is basically the actual terminal device itself. The mobile termination has the

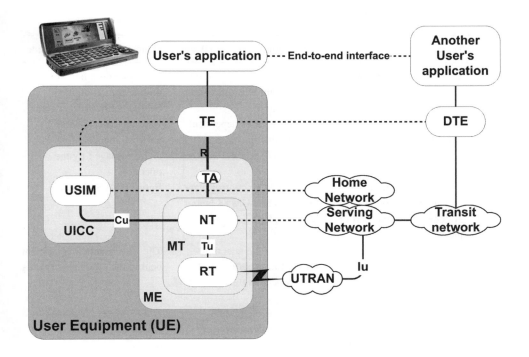

Figure 6.1.   User equipment reference architecture

ability to change its location within the access network or move to a coverage area of another access implementing the same access technology. Mobile termination also terminates the services of the UMTS network systems.

The Network Termination (NT) functional group of the MT is the core network dependent part of the MT. The NT uses non-access stratum protocols for mobility management (MM/GMM) and communication management (CC/SM). Therefore the NT may be seen as the terminal from the pure core network point of view. The non-access stratum protocols are presented in Chapter 9.

The Radio Termination (RT) functional group of the MT is related to the radio access network only. The RT contains the functions common to all services using the same RT-specific radio access technology. The RT uses access stratum protocols like Medium Access Control (MAC), Radio Link Control (RLC) and Radio Resource Control (RRC) on top of the physical radio connection. Therefore, the RT is seen as the terminal from the UTRAN point of view. The MAC, the RLC and the RRC protocols are also presented in Chapter 9.

In this book the terminal architecture is described by using the functional groups described above. A summary of their dependencies on the network side services and subsystems can be found in the Table 6.1. The allocation of main functional entities to the above functional groups clarifies the nature of the functional groups in Figure 6.2.

Strictly speaking, only the final "logical" counterparts in the Core Network (CN) are mentioned here. In some cases the actual protocols may terminate in another network element

Table 6.1 Summary of functional groups in the UE

	Dependencies	Termination on terminal side	Counterpart in network side
UE	User application or user interface independent part of user's telecom facilities	Terminates the application independent telecom system between users	Corresponding fixed or mobile user equipment behind the transit networks
USIM	User's subscription dependent part of UE	Terminates user dependent control functions in principle within his/her home network	Basically user's home network registers (e.g. HLR, AuC) managed by the operator
ME	User's subscription independent, mobile system dependent, part of the UE	Terminates all control plane functions and UMTS bearer for user plane	Entire UMTS network
TE	Telecom service platform dependent, basically mobile system independent, part of UE	Terminates the telecom services transported by the UMTS bearers	The peer terminal equipment behind the external networks
MT	UMTS system dependent, basically telecom service independent, part of ME	Terminates the UMTS network system services	UMTS system managed by active access (visited) operator
NT	Core network dependent, radio access network independent, part of MT	Terminates the UMTS core network services	UMTS CN
RT	Radio access technology dependent, core network independent, part of MT	Terminates the UTRAN services	UTRAN

for optimisation reasons. For example, the mobility management control protocol between the USIM and the HLR often terminates in the MSC/VLR or the SGSN in visited network, because the actual information from the HLR is often duplicated there.

Although the terminal and network side architecture differ from each other some basically corresponding interfaces can be identified on both sides. Naturally, the used radio interface Uu is exactly the same.

The Tu reference point (see Figure 6.1) connects the UTRAN and the CN specific parts together in the terminal side as the Iu does on network side. For performance reasons the Tu reference point is in practise proprietary and embedded inside the UE hardware implementation. The corresponding Iu is in turn standardised interface because there may be different vendors for the UTRAN and CN network devices.

The Cu interface corresponds to the D, C, Gr and Gc interfaces in the CN that connects switching (MSC and GMSC) or routing (SGSN and GGSN) serving network elements to register network elements (HLR, AuC) in the home network. These interfaces are standardised in both sides. In the terminal side because there is an interface between operators and mobile equipment vendors in the Cu. In the network side because there are interfaces between the home and visited operators.

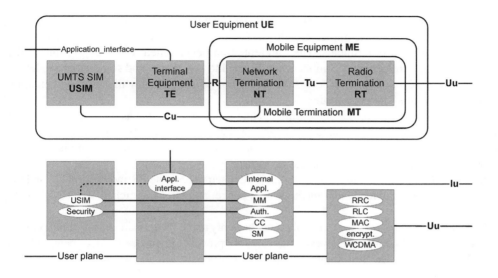

Figure 6.2.   Allocation of main functional entities in a UE

6.2. Differentiation of Terminals

The UMTS terminal has very diverse requirements and these are not necessarily in line with each other as illustrated in Figure 6.3. This increases the complexity of the equipment and

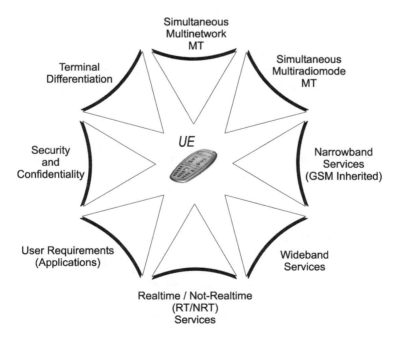

Figure 6.3.   Factors affecting the UE

will have its effect on the equipment price. If the price of the terminal is too high it in turn increases risks in network implementations. These together with many other factors will cause differentiation in terminals. The differentiation development is already visible in the GSM market, where certain MS models are directed to consumers and some other models to business customers and so on. It seems that in 3G this differentiation will cause segmentation in the terminal markets even further.

A remarkable feature of the 3G terminal is that it has, even in its basic form, two different kind of serving network domains available, i.e. Packet Switched (PS) domain for packet services and Circuit Switched (CS) domain for circuit services. From the terminal point of view, it results in three different operating modes:

In the *PS/CS operation mode* the terminal is attached to both the PS and CS domain and the terminal is able to provide *simultaneously* both packet and circuit switched services via both domains.

In the *PS operation mode* the terminal is attached only to the PS domain and may only provide services over the PS domain. However, this does not prevent circuit-like services being offered over the PS domain. The best-known example of an originally circuit-switched service that can also be implemented as a packet-switched service is the Voice over IP (VoIP).

In the *CS operation mode* the terminal is attached only to the CS domain and may only provide services over the CS domain. However, this does not prevent packet-like services being offered over the CS domain. It is even possible that some real time packet services especially with high QoS requirements are more practically provided over the CS domain with allocated constant bit rate although they may be packet switched in external networks.

In the PS/CS operation mode the terminal may have an opportunity to optimise location update procedures if the network supports the optional Gs interface between MSC/VLR and SGSN for this purpose. In this case the terminal may choose according to its capabilities whether to use combined or separate update procedures. In the combined location update procedure the terminal informs only the SGSN about a new location and the SGSN informs then the MSC/VLR about the location area update.

The separation of radio termination and network termination functions in mobile equipment allows classification of the user equipment according to the MT's capability to use several access and core network technologies. The following presents basic alternatives, which theoretically are possible.

- *Single radiomode MT* can utilise only one type of radio interface for user traffic. In UMTS, the preferred radio interface at least in the beginning is WCDMA-FDD, which is therefore a typical example of a single radiomode MT.
- *Multiradiomode MT* can use several radio terminations for user traffic. One interesting case in this category is a GSM/UMTS dualmode terminal, for which the interoperation is well defined in the 3GPP specifications. This kind of terminal enables use of 2G services when outside the WCDMA radio network's coverage. Another implementation may be a UMTS terminal with UTRAN and GERAN access capability.
- *Single network MT* can utilise only one type of core network. A UMTS terminal, which is able to use at least one of the operation modes: PS, CS or PS/CS is an example of a single network MT.
- *Multinetwork MT* is able to use several core networks. Besides the UMTS core network a typical terminal of this kind also supports GSM NSS.

As stated earlier, all of these alternatives offer the basis for terminal implementation, but some of the alternatives may not be commercially attractive due to complicated implementation and/or high costs. On the other hand, UMTS is introduced to a quite mature cellular market, where plenty of 2G services are already available. This in turn creates pressure that UMTS terminals must be able to provide also the existing 2G services from the very beginning.

The first UMTS networks will be implemented on top of GSM and this demands that the UMTS terminals are able to roam in GSM networks. This is also required by the network operators since a terminal that can use both GSM and WCDMA radio access is more attractive and usable benefiting from both access network infrastructures. The circuit switched voice service is especially causing this demand because the subscribers expect their voice call not to be dropped due to possible limited coverage in the first UMTS network installations. In this respect the multiradiomode MT is a way to expand the network coverage.

Thus, typical UMTS terminals will be from the beginning multiradiomode and multinetwork MTs.

Such a UMTS terminal will provide a platform for a very diverse set of services. On the other hand, the services the users require also vary a lot: one may be happy with a conventional voice service and is not interested in sophisticated streaming-type packet switched services. Someone else may see packet switched services as necessary. If the terminal has abilities to handle all kinds of services, its implementation will be very complex and expensive. Also some subscribers may not be happy to pay for additional features they do not consider to be relevant or necessary.

Based on this, we present a rough vision aiming to illustrate a possible segmentation of the UMTS terminals. Please note that this vision does not represents the view of any manufacturer nor specifications, its only purpose is to give an idea of how the different segments and mobile market shattering could happen.

Based on the subscribers and their needs and also taking into account the possibilities the wide band radio access offers one is able to distinguish four main segments describing subscribers, their needs and relevant services.

- *Classic terminal*: this is equivalent to the present cellular phone. It is made to be cheap and thus it contains limited facilities like circuit switched voice access and limited data access with modestly low data rates, which are better than in the present GSM and GPRS. This terminal is able to handle both GSM and WCDMA radio access but not necessarily simultaneously. In other words, it implements "selectable multinetwork MT". This kind of terminal can be considered to be a kind of "GSM extension", its value of use is based on the existing GSM networks and WCDMA access is used occasionally, mainly for packet switched connections.
- *Dual mode*: this terminal type contains both accesses, GSM and WCDMA and it can automatically select the access method based on available coverage and requested service. For instance, voice calls are typically connected through the GSM, but data and multimedia services through the WCDMA. This terminal type is able to utilise the advantages of both accesses and it is also able to perform inter-system handover in both directions. In the case of inter-system handover, the service in use is adapted to the radio access whenever possible. Thus, this terminal type implements "simultaneous multinetwork MT". When the UMTS networks are brought into wide commercial use, these terminals will most probably form the mass market.

- *Multimedia terminal*: this is like the previous terminal but more intelligent from the network point of view. Dual mode does not necessarily handle the UTRAN radio bearers with the best possible manner but the multimedia terminal is able to perform "optimal multiplexing" of the bearer(s) used for multimedia calls. This capacity saving aspect is a very important issue when the UMTS networks evolve. Multimedia terminal is a kind of combination of cellular phone and palm/laptop computer. It contains plenty of applications to handle the multimedia connections and services. This terminal type is not necessarily for the mass market, rather it is directed at business users, at least in the beginning.
- *Special terminals*: these terminals do not necessarily have a "phone presence" like the previous ones. These terminals serve special purposes and they will be integrated together with other equipment. This kind of terminal could be, for instance, located in a car and it could work in co-operation with the vehicle's computer. If the car is stolen this kind of terminal could be "woken up" and it can deliver the vehicle's position information with street addresses by using GPS and the vehicle's computer information. The three previous types implement more or less the CS/PS operation mode. This type is most probably using PS operation mode only and the application areas can be very diverse. There are many visions about intelligent house ware like a refrigerator being able to order more food when required. The special terminal described here could be the integrated communication equipment establishing the connection in this case.

6.3. Terminal Capabilities

Because the life cycle of UMTS networks is expected to span even several decades there will be several kinds of compatible terminals with different kinds of capabilities. An arbitrary terminal and an arbitrary network should therefore be able to somehow negotiate which basic possibilities or alternatives they can use with each other. The UMTS has adopted the GSM-like policy for negotiation. The networks inform their UEs by broadcasting a lot of system information telling the UEs about the network capabilities. The UE knows its own capabilities and informs the UTRAN or the CN.

The basic information set about the capabilities of a UE is called *mobile station classmark*. During evolution in the mobile system technology the number of alternatives has been growing and therefore also the classmark concept has been updated in a backwards-compatible way. The size of the original smallest classmark 1 used in early GSM was 2 octets (1 octet is 8 bits). Classmark 2 was 5 octets and classmark 3 may have a maximum size of 14 octets. Classmarks 1 and 2 are used also in GSM. In UMTS classmark 2 can be characterised as "CN classmark" and classmark 3 as "RAN classmark". Which version of classmark is actually needed depends on the actual procedure. Typically information about the basic capabilities of a UE in the mobile station classmark 3 are:

- Available WCDMA modes like FDD or TDD
- Dual mode capabilities, i.e. support for different variants of GSM systems with supported frequency bands and other special features
- Available encryption algorithms
- Properties of measurement functions in the UE, such as availability of extended measure-

ment capabilities and the time needed for the MT to switch from one radio channel to another for performing a neighbour cell measurement

- Ability to use positioning methods and different kind of positioning method are supported
- Ability to use universal character set 2, i.e. 16 bit character coding standard, known also as ISO/IEC10646 or Unicode, instead of a default 7-bit GSM character set in short message services

Besides the classmarks there is also another similar kind of information element for describing the entirely radio interface specific properties of the UE. This information element is called mobile station radio access capability.

6.4. UMTS Subscription

Like GSM, UMTS networks separate the subscription from the ME. The subscription-specific information set is called a USIM (Figure 6.4). The USIM is also called a "Service Identity Module" because the services actually follow SIM card identification information in every respect. The corresponding information is originally stored in the Home Location Register (HLR) of the home network of the subscriber.

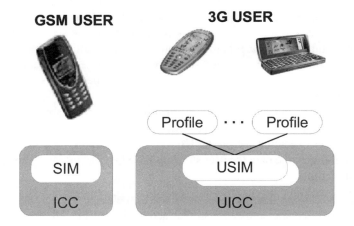

Figure 6.4. Subscriber identity module in GSM and 3G

In GSM the SIM is on a removable ICC (Integrated Circuit Card) and it is the physical place to store subscriber and possible service information. In UMTS the removable physical data storage is UICC and this contains the USIM with service information and identities. The USIM is accessed through profiles, which are "filters" defining how, for instance, stored information is shown to the user. These profiles are subject to change: the user is able to change profile settings and the network can make some changes to profile information, too.

One USIM may contain numerous profiles each of them meant for certain purposes. Let's suppose that a user has two UMTS terminals. One of them is of the "classic type' (please refer to the terminal segmentation issues presented in Section 6.2) and the other is "multimedia terminal". When a user inserts the USIM into either of his/her terminals the subscription is the same but the terminal shows information in a different way. A different profile is used for the same subscription depending on the TE used. For example, through the "multimedia terminal" the subscriber is able to gain access to the information (e.g. picture archive), which is not available through the "classic terminal".

On the network side these profiles are handled by the Mobile Executing Environment (MExE). Another functionality making the USIM information accessible to the TE applications is the USIM Application Toolkit (USAT). The clear difference between a GSM SIM and USIM is that USIM is, as per default, downloadable and its information is accessible and updateable through the radio path.

USIM contains basically five types of data:

- *Administrative*: this is fixed type data assigned by the USIM manufacturer and service provider/operator, such as key values for security algorithms, IMSI and access class information.
- *Temporary network data*: this is mainly mobility management information such as current location area ID, TMSI and calculated ciphering key value(s).
- *Service related data*: this contains information about availability or permissibility of different services and their internal data. A service is available when the subscription has the capability and the permission to support the service. A service not available means, that the service shall not be used by the USIM user, even if the UE had the capability to support the service. For example the USIM may contain, if the operator allows, services such as a local phone book of the user; mobile subscriber ISDN number; fixed dialling numbers; service dialling numbers; barred dialling numbers; outgoing and incoming call information; storage; status reports and service parameters for short message; advice about charging; user and operator controlled PLMN selector with access technology; co-operative network list, etc.
- *Applications*: USIM may store small applications needed for specific services. These applications may implemented, e.g. as Java applets, which are downloaded and stored in USIM for later execution within the UE.
- *Personal*: this covers the data the user stores in the SIM, for instance, short messages and abbreviated dialling.

From these five classes, the first three are fixed in their size and format since they must appear exactly the same way and cause similar actions in any TE. The fourth class, applications, is somewhat undefined. It can be considered as memory but right now there are no exact ideas on how large this area should be. The fifth class has in principle a fixed format but the size varies per operator and subscription basis: some USIM cards are configured to reserve more memory for short messages and abbreviated dialling storage than others.

6.5. User Interface

The 3GPP allows free hands to implement the user interface of the UMTS terminal (Figure 6.5). This arrangement aims to provide cost effective and creative solutions for the terminal user interfaces.

The user interface of UMTS terminals may or may not follow the "traditional" layout used in GSM terminals. The implementation of the user interface depends completely on the terminal manufacturer. It is however very probable that something similar to a normal numeric keypad will be presented in one form or another.

Figure 6.5. Conceptual UMTS terminals

The requirements on the physical layout of input and output features are kept to a minimum in order to allow for differentiated types of UE and to ease the introduction of future developments. However, since the requirements on the physical input features are minimal, the control procedures may differ between UE depending on the solution of the manufacturers. The common denominator between these requirements is that the same logical actions have to be taken by the user. That is, the user has to provide the same information for the call control and signalling no matter what the method is. This is also valid if an automatic device is used for carrying out the same actions.

There are some applicable and mandatory functions the terminal must fulfil, such as, "Accept", "Select", "Send", "Indication" and "End". These functions are essential in order to manage mobile originated and mobile terminated calls and supplementary services. These functions can be implemented in any suitable way, i.e. traditional numeric keypad, voice control, etc. The main issue here is that the user must have access to all of these functions.

"Accept" functionality is used to accept a mobile terminated call. "Select" functionality is used when entering information. The physical means of entering the characters 09, + , * and # (i.e. the "Select" function) may be a keypad, voice detection device, data terminal equipment or other, but there must be a means to enter this information.

"Send" functionality takes care of sending the entered information (called subscriber number, for instance) to the network. "Indication" functionality is used to give all kinds of call progress indications. "End" functionality is used to terminate or disconnect a call. Either party involved in the call may cause the execution of the "End" functionality or, in addition to this, the "End" functionality could be activated due to system level reasons like, for example, loss of coverage or invalidation of payment.

7

Services in the UMTS Environment

In this chapter we present service related matters of UMTS networks. At first, we discuss the commercial model in general. The next main topic is the bearer/Quality of Service (QoS) architecture of UMTS and its effect on the system. The third item is service capabilities (service platforms) and their evolution. Finally, we present some aspects of services and service concepts. Especially we will pay attention to positioning a service platform and service concept since it is expected to be a rapidly growing business area employing a number of new end user services referred to as location based services.

7.1. Services and the Commercial Model

In the days of first generation (1G) the networks were mainly established to offer one service; that is speech. Due to the regional standardisation processes and incompatibility, those systems did not support the idea of having third parties providing services or service contents. With second generation (2G) networks the concept was more open to this development and some commercial bodies focusing their business on services and content provision came to market.

Figure 7.1 illustrates how services, their creation and content provision are located in UMTS. In Chapter 1 we described the "rule of cost" related to these layers in Figure 7.1; the costs are equally distributed on every layer but they have a different nature. The lower the layer is the more hardware related the costs are. We have presented the network element layer in Chapters 4 and 5 and, referring to the bearer/QoS architecture, also clarified some viewpoints of how the service requirements should be handled. The reason for high investments is actually QoS, since the system must be able to handle end-to-end service requirements in all levels described in Figure 7.1. For QoS, we study some additional viewpoints in Section 7.2.

The layered structure presented in Figure 7.1 brings new aspects on the system level. Maybe the most important one is security and its handling. The UMTS commercial model gives possibilities to separate some commercial roles. The bodies acting in these roles have to transfer some sensitive information amongst them and this has a remarkable effect on the security mechanisms. Security in UMTS is discussed in Chapter 8.

The business or value chain in UMTS consists of four parties, end user, carrier provider, service provider and content provider. This separation is only logical, i.e. one or more parties could be the same commercial body but the main point here is that this kind of separation

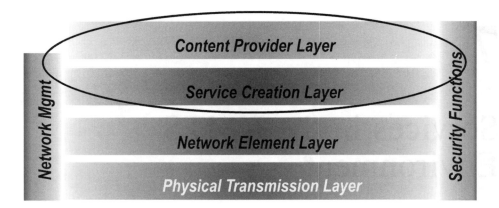

Figure 7.1.   Services and their position

presented in Figure 7.2 is enabled throughout the UMTS system. At the same time with UMTS the Internet world has changed the service methodology a lot. On the Internet side, one maintains the physical site offering the platform for a service. Someone else provides the service itself, this is, prepares HyperText Markup Language (HTML)–pages and other items the service requires. An essential issue is then the content, which again may be provided by yet another player.

Because UMTS partially is the same as "wireless Internet" the UMTS network must be adapted to this commercial model coming from the Internet business. The Third Generation Partnership Project (3GPP) specifications actually aim to define vertical interfaces in the business/value chain. For example, the interface between the end user and carrier provider is the radio interface for UTRA (Uu) interface, which is specified in detail to allow inter-operability. The other interfaces are not so obvious and they vary depending on case but they still exist.

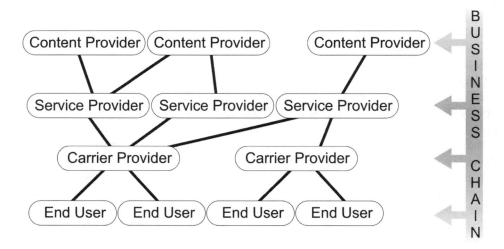

Figure 7.2.   3G services and business/value chain

Carrier provider is the party maintaining both access and Core Network (CN) (in most of the cases), so we are talking only physical network here. The service provider is the one providing the service, which becomes apparent when one notices that a UMTS subscription is also a service. Hence a conventional cellular network operator is actually a service provider from this viewpoint. Content provider is a party creating and providing the end user services. Very typical example of a content provider is a stock exchange giving information about its exchange rates.

In real life, these roles are mixed up, i.e. the service provider and carrier provider are often the same company and this company also makes content provision at least to some extent. This kind of structure may not be the optimal solution in sense of business and the business chain thinking presented in Figure 7.2 may offer attractive possibilities to minimise costs and boost the end user service creation. Naturally this solution does not come without problems: a very obvious difficulty will be the pricing; how the end user is able to determine the total price of the service because each part of the business chain makes its own pricing.

For now, one might think: "Fine model, but what are those separate bodies selling then?" To clarify this, let us look for the UMTS network capabilities from service point of view. To put it simple, a UMTS network as a platform must be able to provide the following bit rates in its coverage area.

As can be seen from Table 7.1, the minimum requirement within the coverage area is 144 kb/s for both circuit and packet switched connections. This bit rate defines the UMTS bearer bandwidth according to the certain probability factor that the carrier provider has used when dimensioning the network. The bit rates defined in Table 7.1 are actually products of the carrier provider and the carrier provider has committed to produce these as cost effectively as possible and maintaining the quality level the business expectations define. The bit rates indicated in Table 7.1 can be understood as target values. In the first UMTS network installations the values indicated in the third row are not readily available due to technical reasons. On the other hand, the other bit rates in Table 7.1 may well be achieved throughout the whole planned coverage area. The question here is, is it affordable or not. The Wideband Code Division Multiple Access (WCDMA)-FDD radio access provides the indicated bit rates but the longer the distance to the BS antenna is, the smaller bit rate will be available. In order to make a successful compromise with investments, coverage and bandwidth, the minimum requirement for the bit rate is around 64 kb/s instead of the 144 kb/s indicated in Table 7.1.

From the carrier provider point of view, it does not matter what kind of information the allocated UMTS bearers carry; it is service provider's business. The real content of the service is in turn coming from the content providers and end users.

Table 7.1 UMTS bit rates

Circuit switched bit rate (kb/s)	Packet switched bit rate	Coverage type
144 kb/s	144 kb/s (peak)	Basic coverage, rural/suburban, fast moving vehicles, outdoor
384 kb/s	384 kb/s (peak)	Extended coverage, urban, moving vehicles, outdoor
2 Mb/s	2 Mb/s (peak)	Hot spot areas, urban, centre, walking speeds, indoor

As a conclusion, the specified bit rates are just edges, and the UMTS network itself is not interested in the information it carries through itself. From the service point of view, the bit rate value does not guarantee anything; the more important issue is the QoS in the UMTS bearers and how it can be utilised and verified.

7.2. QoS Architecture

From the end user point of view, the UMTS network is a network for services. In this respect the technology itself is not the most important but it is an enabler providing QoS so that the end users can be satisfied with the end-to-end services they use. To express this more technologically, the end-to-end services are carried over the UTMS network with bearers. Bearer is a service providing QoS between two defined points. Since the UMTS network structure contains many system levels having their own QoS properties (for example, radio path), the QoS is handled in many levels taking into account the special characteristics related to that level.

In Chapters 3–5 we presented some parts of this architecture in more detail and illustrated what kind of mechanisms have effect on the QoS on those levels. In this Chapter we present some viewpoints relating to the end-to-end service. It should be noted that the term QoS covers both connection control and user data flows.

Based on the bearer service division presented in Figure 7.3, the QoS can be guaranteed well *within* the UMTS network. On the other hand, the QoS problems will probably relate to the item "external bearer service", which aims to cover QoS in connections to the other networks. The QoS mechanisms outside the UMTS network are not in scope of 3GPP specifications and in this respect some problems may occur.

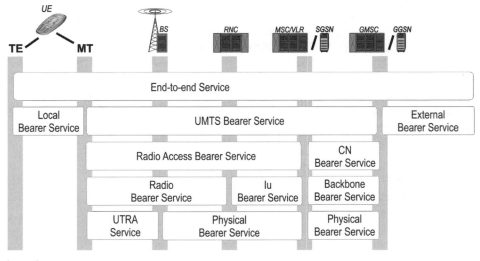

Figure 7.3.   Bearer/QoS architecture in UMTS

Referring to Figure 7.3, the end-to-end service sets the requirement for QoS. These requirements are then mapped to the next level, which in turn performs QoS mapping for the next level and so on. As a result, the UMTS network forms a connection through itself fulfilling the original QoS requirements. To make this mapping possible, the QoS requirements are classified.

7.2.1. QoS Classes in UMTS

UMTS QoS classes (also known as traffic classes) are defined keeping in mind that the classification must be simple. The following four leading principles in this respect can be applied:

- The QoS classes must allow efficient use of radio capacity (UMTS Terrestrial Radio Access (UTRA) service).
- The CN and UMTS Terrestrial Radio Access Network (UTRAN) must be able to evolve independently (division of radio bearer service and Iu bearer service).
- UMTS network must be able to evolve independently from its surrounding networks. On the other hand, the backward compatibility mechanisms must be present (UMTS bearer service and external bearer service division).
- The operator must be able to utilise existing transmission technology within UMTS system in a cost effective way (physical and backbone bearer service).

From the end user point of view, the impression of the connection quality is very often related to the delay experienced on the connection. Due to this, the connection delay is the main separating attribute between the UMTS QoS classes. The other main factors are guaranteed bit rate (bandwidth) and nature of traffic (symmetric/asymmetric).

As defined, UMTS QoS classes are as follows:

- *Conversational class:* minimum fixed delay, no buffering, symmetric traffic, guaranteed bit rate
- *Streaming class:* minimum variable delay, buffering allowed, asymmetric traffic, guaranteed bit rate
- *Interactive class:* moderate variable delay, buffering allowed, asymmetric traffic, no guaranteed bit rate
- *Background class:* big variable delay, buffering allowed, asymmetric traffic, no guaranteed bit rate

The conversational class is the most demanding QoS class meant for Real Time (RT) traffic, for instance, voice calls. Actually all the services using CN CS domain have conversational class on the *UMTS bearer service* level. On the end-to-end service level the user may still experience quality problems if the external bearer service and the other end of the connection are not able to guarantee QoS. Conversational class may also be used in the "overkilling" manner, for example when a user likes to perform web surfing over circuit switched connection (this equals to normal circuit switched data call performed over, for example, a GSM network). The connection itself is capable of doing continuous symmetric bit flow through the UMTS network. The experienced QoS for the web surfing session, however, may still not be satisfactory to the user, since the flow of information is not RT due to overload in the web server or in the external networks involved. As a conclusion, in this case the end user had QoS Conversational class service in use at the UMTS bearer service level, but the end-to-end service is still something else.

The streaming class does not set so tight limits for the delay; delay may vary during the connection and thus the information may be buffered in the network. Typical streaming class services will be asymmetric in nature; for example, the user may download music from the network.

Interactive and background class are meant for end-to-end services not sensitive for delays. A typical interactive class service could be, for instance, Wireless Application Protocol (WAP) service; the user sends a request and the network responses when there are free resources to respond. This is, the delay between response and request may vary and the information to be delivered to the user may be buffered in order to optimise the networks performance and capacity.

The most important QoS parameters in UMTS are:

- Maximum bit rate (kb/s)
- Guaranteed bit rate (kb/s)
- Allowed transfer delay (ms) and
- Whether the requested QoS class is negotiable or not.

The maximum bit rate defines the maximum bit rate the UMTS bearer may use when delivering information between the end-points of the UMTS bearer. Guaranteed bit rate defines the bit rate the UMTS bearer *must* carry between the end-points of the UMTS bearer. Transfer delay, as explained earlier, is the main distinguishing factor between the UMTS QoS classes. Also it should be noted that the QoS class for the end-to-end service requests might be negotiable or not. For example, in the case of circuit switched speech call the QoS class must be conversational and there is no room for negotiation. On the other hand, packet data services could be used within various QoS classes.

Figure 7.4 shows the rough principle of how different bearers are managed in case the end-to-end service in User Equipment (UE) is initiated. First, the Communication Management (CM) entity in UE classifies the requested service based on the service class criteria. After that the UE initiates the UMTS bearer establishment over UTRAN. This request is carried within the message named "UE initial message" through UTRAN in the payload of a Radio Resource Control (RRC) message. From the UTRAN to the CN this message is carried in a Radio Access Network Application Part (RANAP) message. At the CN side, the end-to-end service require-ments are finally mapped onto a UMTS bearer based on the defined QoS attributes. This mapping is performed by the communication management task in CN. The end-to-end service requirements are investigated against the QoS parameters indicated in Figure 7.4.

The related CN domain also checks the UMTS bearer requirements and starts Radio Access Bearer (RAB) allocation through UTRAN. RAB assignment request is investigated by the Radio Resource Management (RRM) Admission Control (AC) algorithm, which checks whether the RB for this transaction can be established with the requested QoS para-meters. If RBs are not available and the requested QoS is negotiable the QoS parameters are renegotiated between UTRAN and CN. If the QoS is non-negotiable and the requested QoS of the RAB cannot be satisfied, the request is rejected or queued. If AC in the Radio Network Controller (RNC) allows, the RB is established with given QoS parameters and then the Iu bearer between UTRAN and CN is also set up. Now the UMTS bearer is ready to carry data flows according to the end-to-end QoS requirements.

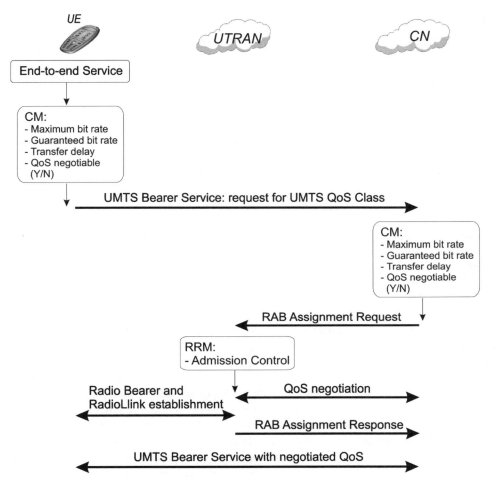

Figure 7.4.   Rough principle of bearer management

The principle described above and in Figure 7.4 is mainly valid for both circuit switched and packet switched traffic. Roughly, the difference between these traffic types is that, in the case of packet switched connections the UMTS bearer is released when there is nothing to transfer but the addressing and QoS information is stored in Packet Data Protocol (PDP) context (refer to Chapter 5). In the case of circuit switched connection the UMTS bearer exists until the connection is released because it always uses dedicated radio resources. Also, the addressing and QoS information is not valid when the circuit switched connection has been cut.

Tables 7.2 and 7.3 give some information about the QoS attributes the system uses when mapping end-to-end service requirements to the bearer services. In addition to those values listed in Tables 7.2 and 7.3, there are numerous other attributes indicating, for instance, allowed error ratios in different cases, priority handling indication and length of delivered Service Data Units (SDU).

Table 7.2   UMTS bearer service attributes (partial and indicatory)

	Conversational	Streaming	Interactive	Background
Maximum bit rate (kb/s)	Less than 2048	Less than 2048	Less than 2048	Less than 2048
Guaranteed bit rate (kb/s)	Less than 2048	Less than 2048	N/A	N/A
Symmetry	Symmetric	Asymmetric	Asymmetric	Asymmetric
Transfer delay (ms)	100–250	250–seconds	N/A	N/A

Table 7.3   UMTS radio access bearer service attributes (partial and indicatory)

	Conversational	Streaming	Interactive	Background
Maximum bit rate (kb/s)	Less than 2048	Less than 2048	Less than 2048	Less than 2048
Guaranteed bit rate (kb/s)	Less than 2048	Less than 2048	N/A	N/A
Symmetry	Symmetric	Asymmetric	Asymmetric	Asymmetric
Transfer delay (ms)	80–250	250–seconds	N/A	N/A

The end-to-end service requirement and their mapping to the UMTS bearer are defined in service-specific 3GPP specifications. These specifications (Series 22) define QoS and performance requirements for each specific service.

As can be seen from Tables 7.2 and 7.3, the difference between QoS classes is mainly due to transfer delay. Because the delay is such an important factor, the aim is to absolutely guarantee that the transfer delay is never exceeded. This is why the RAB has in conversational QoS class tighter transfer delay limit than on the UMTS bearer level. Also it should be noted that the minimum values for transfer delay are fixed, but the other end of the value range depends on UTRAN performance; the better UTRAN performance, the smaller transfer delays as average. It is assumed here that the transmission system used is able to provide adequate bandwidth.

So far, we have discussed UMTS bearer and RAB service. These are the main bearers, but the mapping continues further, as was indicated in Figure 7.4, where the RRM involvement in RB was shown. Within the CN the CN bearer and Iu bearer are the points of interest. These bearer services describe quite closely the physical media to be used for the UMTS bearer and there are actually three options to guarantee QoS on this level:

1. If the transport network connecting CN nodes and UTRAN together uses Asynchronous Transfer Mode (ATM) as transport, the CN bearer service attributes are mapped to ATM traffic classes. This is the "de facto default" in 3GPP R99 implementation of UMTS. As time goes by and the network is evolved to 3GPP R4/5 the Internet Protocol (IP) will be used as transport.
2. IP can be implemented on top of ATM; here we are talking about IP/ATM. In this case, the CN bearer and Iu bearer QoS attributes can be mapped either to ATM traffic classes or the operator may use IP-specific QoS mechanisms like Differentiated Services (DiffServ). This kind of hybrid solution guaranteeing QoS works well but it could be quite complex to manage since there are various protocols in different levels used for the QoS management.

3. The third option is to use IP as transport over physical media offering enough bandwidth. In this case the CN and Iu bearer QoS attributes are handled with IP-specific QoS mechanisms like DiffServ.

Actually it is the carrier provider's choice whether to use ATM or IP as transport even in 3GPP R99 but ATM can be seen as the default alternative.

7.2.2. QoS Classes – a Provocative Approach

In 1G and 2G networks the connection has, so far, been mainly circuit switched in nature. This kind of connection has its advantages and disadvantages. An advantage, circuit switched connection is very stable when allocated: it uses fixed bit rates in both directions, symmetric and providing stable media for any service using the connection. A disadvantage, circuit switched connection easily wastes radio resources if the service used in the connection does not require fixed bit rates and symmetric type of connection. In UMTS network the majority of the traffic is expected to be packet switched in nature and packet switched connections bring new dimensions into the picture; the connections are not necessarily using fixed bit rates any more and, in addition, the bit rates may differ according to the transmission direction. The changes described in the previous sentence are classified and system behaviour is defined against these classes especially with packet switched connections. This classification all together is called QoS.

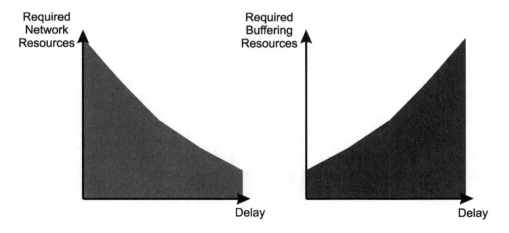

Figure 7.5.   QoS and affecting factors (provocative)

The QoS is relatively a subjective issue because it originally describes the "human factor" concerning the quality of the connection. Because of this, the major issue in the QoS is the delay; how long it is and how variable it is in nature. The QoS is the better the shorter of the delay of the connection is in general. This creates the impression of good quality connection as far as human beings are concerned.

When studying the QoS from the network point of view it has several other affecting factors presented in Figure 7.5. First of all, minimising the delay (in average) requires remarkable amount of network resources. In this context network resources mean, for example, transmission bandwidth allocated for the connection. The longer delay allowed, the less network resources in average are required. This dependency can be described using the graph on the left hand side of Figure 7.5. The shape of the graph can, in theory, universally be calculated but in practise it is operator dependent due to the existing investments and other related issues.

Another viewpoint is then the buffering of the bit streams or information packets. At first glance buffering itself sounds a good idea: by momentarily storing the information flow within the network some valuable network resources could be minimised. This is true, but if the delays allowed in the network are very long, the amount of the buffered information increases. This in turn creates a "snowball" effect; the need for buffering resources increase rapidly as a function of delay. Also the end user may not be happy with the network quality and they may quit subscriptions easily. When the graphs presented in Figure 7.5 are put together, the situation looks as follows:

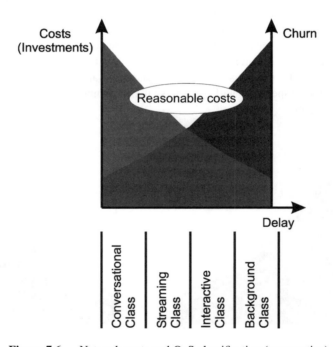

Figure 7.6.   Network costs and QoS classification (provocative)

The common factor for the network resources and buffering needs is money, i.e. the costs and investments. If these are observed as a function of delay, a very rough and provocative Figure 7.6 shows the situation. The delay is classified with QoS and both very good and very bad QoS are actually relatively expensive options. In this respect, the offered QoS is one component of UMTS networks being subject to optimisation.

A very high quality network has its costs on the investment side and the quality offered could be even "too good". A very bad quality network has its costs partially on the invest-ment side and partially in lost subscriptions (Churn). In the middle of these two ultimate options the quality offered is good enough and subscriptions are not lost to competitors. This combination minimises infrastructure investments when compared to the gained revenues. This optimal point naturally varies from operator and country basis and it also partially consists of the end user expectations.

The same issue also can be thought of in another way: certain services are more profitable than others. Based on the graph in Figure 7.6 it can be easily seen that the services requiring QoS Streaming or interactive class are as such because the investments made to the network are used optimally. In real life, the carrier provider however, must implement all QoS classes and the optimisation occurs when estimating how widely a certain QoS class is used.

The definition of the cost level and thus the average QoS definition is one of the strategic decisions of the carrier provider and there are also many other aspects involved. The QoS classification is in practise meaningless for the circuit switched connections and thus, in order to start cost optimisation in the long run, the offered services should be packet switched. This kind of thinking is the main reason behind the hurry to convert as many circuit switched services as possible to packet switched services. For instance, Voice over IP (VoIP) is a major point of interest due to the cost saving possibilities it will offer in transmission bandwidth.

7.3. Service Capabilities as UMTS Service Platform

One of the main principles in UMTS is that the network establishing connections is separated as much as possible from the network parts maintaining services. This differentiation creates more commercial potential and openness in the market place. To implement this development, 3GPP R99 introduces network components, which are called service capabilities, also known as service platform. This part of the UMTS network will experience the most dramatic changes as time goes by.

The most important service capabilities already available in the 3GPP R99 (Figure 7.7) could be listed as follows. Please note that the list is not exhaustive:

- WAP server/WAP gateway: this service capability offers a browser for the end users. The browser can be textual (in the beginning), later on it will contain more sophisticated features.
- Positioning servers: this service capability provides other service capabilities with UE position information. Thus, these services are often referred to as location based services.
- Mobile Station Application Execution Environment (MExE): this service capability provides other service capabilities with information about the terminal abilities to handle information.
- UMTS SIM Application Toolkit (USAT): this service capability offers the tools required for SIM card handling.
- Customised Applications for Mobile network Enhanced Logic (CAMEL): this service capability is populated with numerous services the end users, i.e. this is a common platform for all kinds of services the subscribers may use. When the UMTS network evolves from 3GPP R99 to 3GPP R4, the role of CAMEL increases and it will start acting as a "service interconnection point", i.e. it will have a central role in service implementation. One example of this is Virtual Home Environment (VHE), which is described at the end of this chapter.

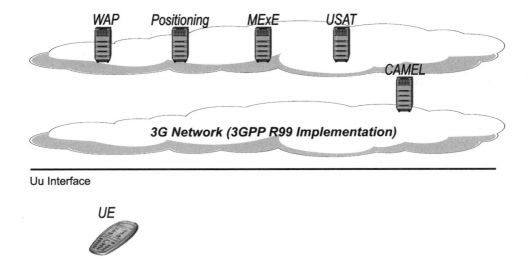

Figure 7.7.   Service capabilities in 3GPP R99 implementation

In 3GPP R99 implementation the service capabilities are somewhat proprietary ones and thus they may be partially vendor-dependent. They do not have complete and common control and, because of this, there is not any universal service creation environment. Figure 7.8 introduces the 3GPP R4/R5 implementation, where the service capabilities are more "uniformed": they have a common, open interface(s) called Open Service Architecture (OSA). The OSA enables the possibility to have a common, centralised point through which the services can be created. This point is called Service Creation Environment (SCE). Through the SCE anyone creating services have access to the service capabilities available under the SCE. Within OSA, the separate service capabilities have Application Programming Interfaces (API), which contain management and control entities for each service capability. Hence, there will be, for example, MExE API and USAT API in the interface between the SCE and related service capability within the OSA system.

In order to effectively use the service capabilities, the CAMEL is, in practise, a mandatory functionality to be included. Since CAMEL provides open protocol interfaces, it is possible to encapsulate service creation and content provision from the CN. Referring to Figure 7.1 in the beginning of this Chapter and value chain presented in Figure 7.2, this structure actually forms a service creation layer and this layer can actually be a commercial body of its own: since it has open interface(s) towards the CN, it does not matter to which CN(s) it is attached. This is a major difference compared to GSM: in GSM the Intelligent Network (IN) based service creation was only valid and available in home network. In UMTS the services are available in every UMTS network if the connection through CAMEL is available and the carrier and service providers have agreed on the services to be used between networks.

The following subsections introduce service capabilities. From these, the positioning is handled in more detail here, since it has not been described in detail in Chapters 4 and 5 and it is expected to be a rapidly growing business area.

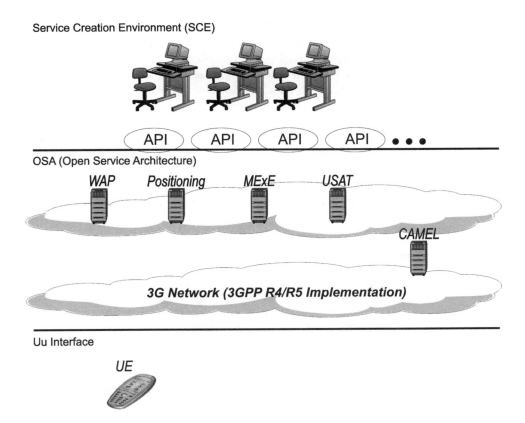

Figure 7.8.   Service capabilities in 3GPP R4/R5 implementation

7.3.1. Service Capability – Wireless Application Protocol (WAP)

Principally, WAP is a way to provide (web) browser for the end user (Figure 7.9). The browser application is located at the terminal and, upon user's generated requests, the browser retrieves information to be downloaded from the network.

When WAP is implemented over a UMTS network, the HTML used in the Internet is not suitable for the mobile environment as such because the HTML format is not exact enough and this may lead to confusion as far as different terminals and their browser implementations are concerned.

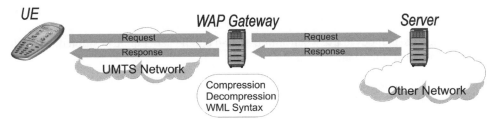

Figure 7.9.   WAP dialog – principle diagram

Instead of HTML the WAP system uses Wireless Markup Language (WML), which is more exact in syntax and format. The WML pages can be stored in a normal Internet server but they are not readable through a normal Internet browser.

The core of the WAP service is a WAP gateway. It is a server able to receive user requests and reformatting them to HyperText Transfer Protocol (http) requests and vice versa. Because used bandwidth is an essential issue in mobile networks the WAP gateway performs data packaging and compression towards the terminals. WAP is a suitable service capability for the services requiring interactive QoS class using request-response model. The delay of the connection is not an issue as such, rather it can be stated that the opening of the connection should be relatively fast.

WAP as such is not in the scope of the 3GPP specifications. For further details about WAP, its existing implementation and future trends, refer to the web address http://www.wapforum.org.

7.3.2. Service Capability – Location Communication Services (LCS)

Mobile positioning is becoming a very potential feature of the future cellular systems. There are two main drivers for this evolution path: advanced position based services and radio system performance optimisation by utilising the position assistance data. Examples of commercial position based applications are fleet management, traffic information management, transportation, nearest services, emergency services, follow me services, etc. On the other hand, position assistance data can be applied for optimising the system performance by utilising it for network planning, RRM, Mobility Management (MM) and so forth.

So far, cellular technology has enabled the end user to get rid of wired communication by supporting user's mobility and various communication services regardless of his/her position. Although the mobility of cellular terminals causes complexity and load within the network it is also becoming an important opportunity in association with the positioning feature of cellular systems.

In this subsection, we first describe the basic principles of different positioning methods that have so far been developed for the mobile networks. Then, we describe those positioning methods that have been specified for the UMTS. We further give an insight into the UMTS positioning architecture as well as the positioning functions allocated to different network elements.

In order to maintain consistency with the 3GPP specification terminology, we use the term "position" as a general concept, but stick to "location" when referring to the 3GPP specifications, in which this system functionality is called LCS.

7.3.2.1. Overview of Positioning Methods

There are various methods to use when aiming to find out the geographic position of a cellular terminal. The most important ones are:

- Cell ID based positioning
- Round-Trip Time (RTT) based positioning
- Time-of-Arrival (TOA) positioning
- Time Difference of Arrival (TDOA) positioning

- Angle of Arrival (AOA) positioning
- Reference Node Based Positioning (RNBP)
- Global Positioning System (GPS)

The basic principles how these methods are used for UE positioning are described next. After that a short summary on the accuracy of different methods is given before turning to the UMTS-specific positioning methods.

Cell ID Based Positioning

The position of the terminal can be estimated using the information on which radio cell the terminal is camping or has recently been camping. The cell related position information consists of either cell ID or cell coverage co-ordinates. By using the cell ID or cell coverage co-ordinates of the fixed known position of the serving BS, the position of the terminal is estimated.

It should be noted that in third generation (3G) specifications, the actual cell ID is seen as radio information and may therefore not be transported to the CN. Instead, the cell ID is mapped into "service area identification", which may be sent to the CN.

Round Trip Time (RTT) Based Positioning

The RTT of the radio signal may also be used in order to increase the accuracy of the position estimates. The calculation of the terminal position may use the RTT measurements of the signal branches from several BSs (Figure 7.10). The RTT is the propagation delay time of the signal travelling from the terminal to the BS and back. Therefore, the distance between the terminal and the BS can be obtained based on the time t and the velocity of the radio wave. Based on this information we have:

$$D = \frac{RTT}{2} c + \varepsilon$$

where D depicts the distance from mobile terminal to the BS, c is speed of electromagnetic wave, which is equal to that of the constant speed of light (3×10^8 m/s) in a free space medium as is the case in cellular system environment. RTT is the round trip time and ε is the measurement error.

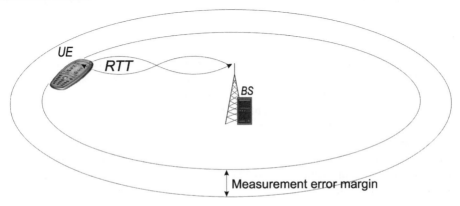

Figure 7.10.   Round Trip Time (RTT) positioning

It should be noted that the distance D is in fact the radius of a circle around the BS. Therefore, for more accurate estimates the location of the mobile terminal can be figured out by using additional RTT measurements from neighbouring BSs, if available. Then, the position estimate is within the intersection of the three neighbouring circles with the BSs located in the centre of those circles. The resulting position estimate accuracy is determined by the error margins of all RTT measurements.

Time of Arrival (TOA) Positioning

In the TOA method the position calculation is based on the propagation delay of the radio signal from the transmitter to the receiver. When there are at least three TOA measurements available, the position of the terminal can be carried out by applying a triangulation technique, minimising the least-square distances between the terminal and corresponding TOA circles. Figure 7.11 shows the principle to estimate the terminal position by measuring the TOA of three neighbouring BS.

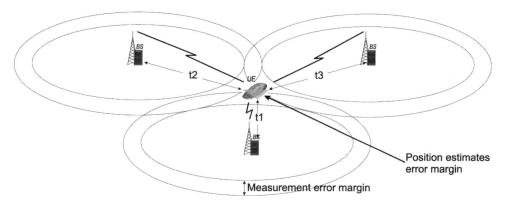

Figure 7.11.   Time of Arrival (TOA) positioning

In the TOA method the distance from the terminal to the BS is proportional to the propagation time, t_i. Thus, if there is Line-of-Sight (LOS) between the handset and BS, the distance between them is given by:

$$D_i = ct_i$$

where c is the speed of an electromagnetic wave and t_i is the delay time for the signal to travel from a BS to the terminal or vice versa. However, practically, there is always some error regarding the TOA measurements due to Non-Line-of-Sight (NLOS), signal fading, reflection and shadowing. Therefore, a measurement error margin should be taken into account when estimating the accuracy of TOA positioning.

As can be seen from Figure 7.11, in order to obtain a unique position estimate of the mobile terminal the TOA of at least three BS is needed. Then, the position of the handset can be estimated by calculating the distance between the terminal and several BSs (TOA circles) in a least square sense. That is:

$$D_i = \sqrt{(X_i - x_m)^2 + (Y_i - y_m)^2 + (Z_i - z_m)^2} + \varepsilon$$

where (X_i, Y_i, Z_i) are the co-ordinates of the neighbouring BSs involved in the positioning process and $(x_m, y_m z_m)$ are the co-ordinates of the terminal to be positioned.

The TOA method would require very accurate BS synchronisation, which may bring drawbacks to unsynchronised cellular systems. The position calculation entity should also be capable of discerning the time difference between the transmitted and the received signal.

Time Difference of Arrival (TDOA) Positioning

The basic principle of the TDOA method is different from that of the TOA method. In the TDOA method, the terminal observes the TDOA of the radio signals from the neighbouring BSs. The unknown terminal position is estimated by processing the TDOA measurements between the terminal and at least three BSs of known co-ordinates. The TDOA measured by the terminal consists of two components:

$$TDOA = RTD + GTD$$

where the Geometric Time Difference (GTD) comes from the geometry, for example propagation delay differences between the handset and the two BSs. GTD is the actual quantity containing information concerning the handset position, since it defines a hyperbola between the two BSs. In addition, in a system like WCDMA, where the BSs are not synchronised, the Relative Time Difference (RTD) must be known. RTD is the transmission difference between the signals of the neighbouring BSs.

Figure 7.12 depicts the principles of the TDOA positioning method. The position of the terminal is calculated based on the TDOA between the serving BS and the neighbouring BSs, which define a hyperbola, the foci of which coincide with the corresponding BS transmitter antenna (co-ordinates of). The position of the terminal can be estimated by utilising the least squares of the distances of the terminal to the hyperbolas, if there are more than two TDOA values available.

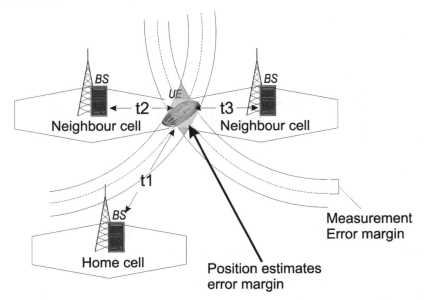

Figure 7.12.   Time Difference of Arrival (TDOA) positioning

Therefore, in the TDOA method, the basic idea is to determine the relative position of the terminal by examining the difference in time at which the signal arrives at the target (terminal or BS), rather than the absolute arrival time. Therefore, if this time difference between mobile terminal and neighbouring BSs is available and there is LOS between the terminal and the BSs, the mobile terminal is located on the hyperbola:

$$D_1 - D_3 = c\Delta t = c^* GTD = \sqrt{(X_i - x_m)^2 + (Y_i - y_m)^2 + (Z_i - z_m)^2}$$

where D_i is the distance from mobile handset to neighbouring BSs and c is the speed of light, Δt is the time difference of arrival of the signal from neighbouring BSs, X_i Y_i and Z_i are the co-ordinates of the neighbouring BSs, and (x_m, y_m, z_m) are the co-ordinates of the mobile terminal. As shown in Figure 7.12, by measuring two TDOA of three different BSs, the position of the terminal can be estimated in the intersection(s) of these hyperbolas. However, to obtain a unique and more accurate estimate there should be at least three BS measurements.

Angle of Arrival Positioning (AOA)

The position of the terminal can also be determined by calculating the intersection of two lines of pilot signal branches, each formed by an angle from the BSs to the mobile terminal. A single measured angle forms pairs of lines and provides the target handset position. There-fore, as shown in Figure 7.13, if there is LOS between the mobile terminal and two BSs and AOA measurements of two neighbouring BSs are available, the mobile terminal will be located in the intersection of the lines defined by the angles of arrival.

Similar to the TOA and TDOA method, the AOA method accuracy can be improved by using more than two measurements. Nevertheless, reflection and diffraction cause severe errors, when AOA is used in an urban environment. Normally the mobile terminal is not able to determine AOA by itself, so in practise the AOA should be measured at the BSs. It is quite complicated and expensive to implement AOA, because several large antennae are needed at the receiver sites. On the other hand, existing diversity antennae could be utilised for this purpose.

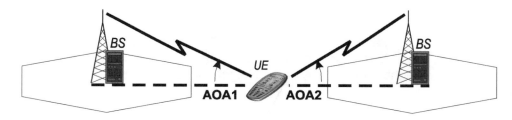

Figure 7.13.   Angle of Arrival (AOA) positioning

Reference Node Based Positioning (RNBP)

The RNBP method is based on the principle that a reference node is chosen to provide auxiliary positioning measurements of the mobile device. The reference node can be movable or fixed and may have relay or repeater functionality. The reference node may be a position-ing service device, a GPS receiver or any device with known position, which can be used as a reference point when determining the position of terminals.

In principle, the RNBP can be utilised with any positioning method such as TDOA, TOA or AOA. The main point is that RNBP may provide enhanced positioning, by utilising additional reference devices in the network. On the other hand, there may be cost drawbacks with RNBPs, since they would have to be placed in a BS site or similar site, and would also require additional antenna. However, the cost drawback can be compensated for, when utilising the otherwise existing network infrastructure in terms of nodes and repeaters in radio networks. In any case, careful network planning would be needed to avoid increased repeater interference levels.

Global Positioning System (GPS)

GPS is a satellite based positioning system, which in many ways offers the best radio navigation aid currently available. GPS is operated by the US Government and is used both for military and public purposes.

GPS can be combined with cellular applications in several ways. The first step is to implement a GPS receiver in the terminal, which offers the same benefits as a standalone

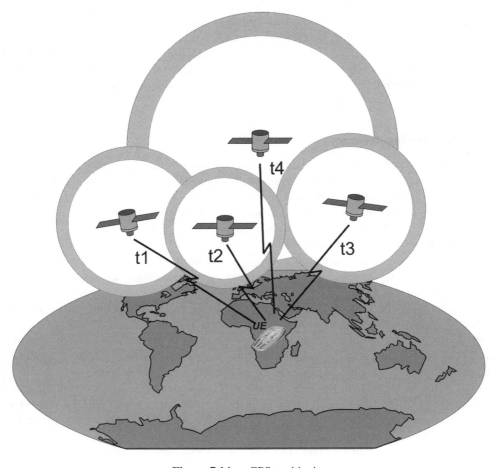

Figure 7.14.   GPS positioning

GPS receiver. The accuracy and speed of the GPS receiver can be radically improved by sending GPS assistance data to the terminal. The assistance data is generated using a GPS reference receiver in the cellular network.

The principle behind GPS is based on the TOA method but due to the accurate timing the implementation of GPS can be quite complicated. GPS requires precise timing within a group of satellites involved in the positioning process by transmitting a spread spectrum signal to the earth on the L-band. A precise clock at the receiver measures the signal's time delay between the satellites and the receiver. This permits calculation of the distance from the receiver to each satellite. In case the receiver observes three satellites its position can be estimated by utilising the triangulation approach. In practice, the clock of the receiver need not be so accurate, if the signals from a fourth satellite are used to correct receiver clock errors. The spheres around the satellites can be schemed based on the signals' travelling time. Having defined three spheres, the receiver's position is calculated as the intersection of these spheres, providing co-ordinates in latitude, longitude, and altitude. Figure 7.14 illustrates GPS based position estimation.

GPS positioning is one of the most accurate positioning system currently in existence, especially when using Differential GPS (DGPS) assistance. However, the GPS receivers normally should have LOS to some four satellites and this sometimes causes problems, especially indoors. Moreover, the GPS receiver has a dedicated satellite antenna, which is typically relatively big and may interfere with the cellular usage. The GPS receiver increases manufacturing costs and power consumption of the mobile terminal. In order to overcome, or at least ease such problems, GPS positioning and GPS assistance have been standardised both for GSM and UMTS.

7.3.2.2. Positioning Accuracy

As was previously described, the main principles of any positioning method, excluding the cell-information based method, are based on the radio signal propagation characteristics, which are not easy to predict. Therefore, the accuracy of cellular positioning methods depend strictly on the cellular environment, availability of LOS, signal measurements, the receiver features, as well as the utilised calculation approach.

The accuracy of the cell ID based method depends strictly on the cell structure of the radio network. In a network structure with macro-cells the position accuracy ranges from a few kilometres to tens of kilometres. In a pico- and micro-cell environment the degree of accuracy will vary from a few to tens of meters.

The accuracy of the TOA method depends on the signal fading, shadowing, and the number of signal branch measurements available for different BSs. It is essential to have LOS to several BSs in order to obtain accurate position estimates of the mobile terminal. Unfortunately this requirement is in contradiction with normal cellular network planning principles, so optimum accuracy cannot be achieved everywhere in a cellular network.

Like in the TOA method, the accuracy of the TDOA method is subject to the inherent characteristics of the radio signal in the cellular network environment and the number of available measurements, especially the NLOS cases. When TDOA is used, however, because only time differences are measured, some errors are common to all BSs and cancel out in the positioning process. Because of this TDOA based methods normally offer better results than TOA based methods, otherwise the preconditions remain similar.

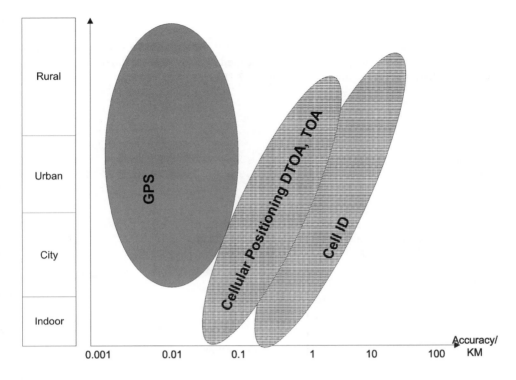

Figure 7.15.   Accuracy of different positioning systems

All in all, there are no reasonable absolute estimates to generalise the accuracy of different positioning methods and thus the positioning accuracy should be evaluated case by case. Nevertheless, the degree of accuracy of different methods may be roughly outlined, as shown in Figure 7.15.

7.3.2.3. Positioning Methods Specified for UMTS

All the positioning methods described above were investigated and discussed in the 3GPP from a standardisation point of view. As a result, three methods were selected for UMTS networks:

- Cell ID based positioning
- Observed Time Difference of Arrival (OTDOA) positioning, which – as the name suggests – is based on the TDOA principle described above
- Assisted GPS positioning

These methods can, in principle, be network based, mobile based, network based mobile assisted or mobile based network assisted. The difference between these variants is whether the position calculation is carried out in the network or in the terminal. Other variants are defined depending on where the assistance data is generated, so the method may be termed network assisted or mobile assisted, respectively. The following subsections give an insight into these positioning methods specified for UMTS.

Cell ID Based Positioning in UMTS

In the cell-ID based method, the SRNC maps the cell ID into the corresponding Service Area Identifier (SAI). Additional operations may be needed if UE is in a soft handover state or in a RRC state, where the cell is not defined (Figure 7.16).

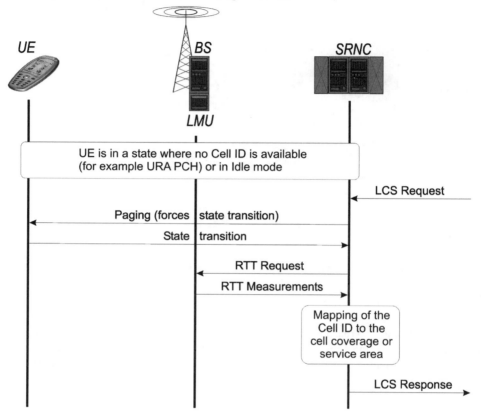

Figure 7.16.   Cell ID based positioning, general procedure

In a soft handover state the UE may have different signal branches connected to different cells, reporting different cell IDs. In this case the cell coverage based position estimate may be improved by combining all available information in the SRNC.

Normally, the cell ID selection is based on the parameters defining the quality of the received signal branches. That is, SRNC selects the cell ID with the strongest signal branch as a reference when determining the UE position estimate. Alternatively, the first cell ID that was generated during connection establishment may be chosen as the cell for position estimation. In addition to these common mechanisms, other criteria have been specified for determining the cell ID during handover.

The cell ID can be determined according to the primary common pilot channel (CPICH in WCDMA-FDD) in the active set of BSs received by the terminal. Also a CPICH that is not included in the active set may become better than the primary CPICH that belongs to the active set and is then used to determine which cell ID to use.

The cell ID based method should support the UE positioning regardless of the RRC state of the connection, that is URA PCH, cell PCH, cell DCH, cell FACH, cell reselection, inter-system modes, as well as an idle mode.

It is not possible to obtain the cell ID if the UE is not in an active state, i.e. if there is no connection between the UE and at least one of the cells. For example, in UMTS the cell ID can be provided only when an RRC connection exists between the UE and at least one BS. Therefore, if the UE is in a mode where the cell identifier cannot be provided, the UE is forced into a state where the cell identifier may be provided. For example, in a URA PCH state the cell ID may not be available, so UE is then forced into a channel state having the cell ID, e.g. cell FACH, in order to determine the cell identifier. The network can also prevent UE from entering the URA update state in order to receive cell ID updates when the UE selects a new cell. When the UE is in idle mode or URA PCH mode, it needs to be paged for positioning purposes by either the CN or UTRAN.

In UMTS, the UTRAN aspects should be hidden from the CN. Because of this principle the cell identifier is mapped to the service area identifier, which is sent from UTRAN to the CN over the Iu interface. The service area could in principle include more than one cell, but currently there is a one-to-one relationship between the cell ID and service area identifier in UTRAN.

Observed Time Difference of Arrival-Idle Period Down Link (OTDOA-IPDL) in UMTS

The second UMTS positioning method Observed Time Difference of Arrival-Idle Period Down Link (OTDOA-IPDL) consists of two parts. Firstly, OTDOA, which is theoretically similar to TDOA described previously. However, in this concept it is emphasised that the UE measures the observed time differences between neighbouring BSs in order to estimate the position of the UE.

Measuring OTDOA and RTD is not very straightforward in WCDMA-FDD, because of the following two reasons:

- In some cases a sufficient number of downlink pilot signals may not be available to be measured by the UE. This may occur when the UE is located so close to the serving BS that its receiver is blocked by the strong BS transmission. This is referred to as the "*hearability effect*".
- The BSs are normally not synchronised in WCDMA-FDD, so the synchronisation difference between BSs, the RTD, should be known or measured, before the position of the terminal can be calculated.

To overcome these problems two possible solutions were investigated, first in the Japanese standardisation bodies and later in the 3GPP, to enable support for UE positioning in WCDMA-FDD networks.

The first solution would be to temporarily increase the transmission power of the neighbouring BSs, to enable the UE to measure them. However, because WCDMA-FDD, like every CDMA based radio technology is interference-limited, the power-up approach was not feasible at all, because it would substantially increase the interference levels in the network.

The other solution is to have the serving BS to cease its transmission for short periods of time in order to reduce the hearability problem. This method is called "Idle Period Down-Link" (IPDL). During an idle period of the serving BS, the UE can measure the signals of neighbouring BSs. During the idle periods also the RTDs can be measured more accurately. The measurement units used for this purpose can be placed in, or close to, BS sites, when IPDL is used.

Assisted GPS for UMTS

There are two basic types of assisted GPS, namely UE based and UE-assisted, which differ in where the actual position calculation is carried out. It should be noted that the UE might also receive GPS assistance data from the network in the UE-assisted GPS method (Figure 7.17).

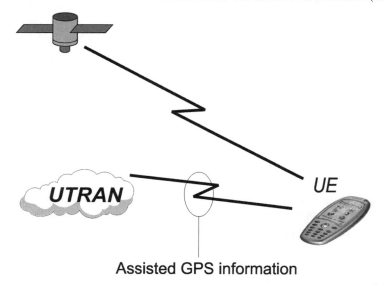

Assisted GPS information

Figure 7.17.   Assisted GPS

In the UE based assisted GPS solution the UE contains a complete GPS receiver, and the position calculation is carried out by the UE. The GPS assistance data sent by the network to UE includes:

- Measurement assistance data, i.e. GPS reference time, visible satellites list, satellite signal Doppler and code phase search window. This data is valid for 2-4 hours.
- Assistance data for position calculation; i.e. reference time, reference position, satellite ephemeris and clock corrections. This data is valid for four hours.

When differential GPS is utilised, also differential correction data is sent to the terminal. This type of assistance data is only valid for about 30 s, but is relevant for quite a large geographical area. It is therefore possible to generate the assistance data using only one centrally located reference receiver.

In the network based UE-assisted GPS solution, the UE only has a reduced GPS receiver. Using this receiver the UE carries out pseudorange measurements and transmits the measurement results to a calculation unit in the network, that carries out the rest of the GPS operation. In this solution the network only sends limited assistance data to the UE, but UE on the other hand has to send all measurement results in the uplink direction. The reference time must also be more accurate in the UE-assisted mode than in the UE based mode.

7.3.2.4. UMTS System Architecture for Location Services (LCS)

The Location Services (LCS) functionality in UMTS is distributed to existing network elements basically in the same way as in GSM. One main new network element, the *Gateway Mobile Location Centre*(GMLC), is added to the overall system architecture to support location services both in GSM and in UMTS.

Figure 7.18 illustrates the overall architecture of LCS in UMTS. The LCS equipment and functionality is located both in CN and UTRAN. The UTRAN contains functionalities responsible for implementing positioning with the above described methods and positioning data collection. In CN the GMLC acts as a connection point through which the positioning data can be delivered to other service capabilities or client applications.

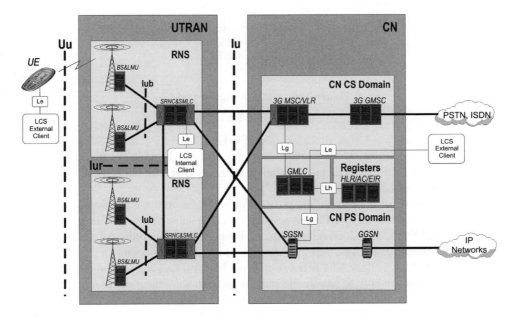

Figure 7.18.   General LCS architecture in UMTS

In GSM, the *Serving Mobile Location Centre*(SMLC) can either be a dedicated physical network element or integrated in the BSC. In UMTS the SMLC functionality is integrated in RNC in the initial specifications, but a standalone SMLC for assisted GPS is also being standardised.

The SMLC in GSM and the serving RNC in UTRAN select the positioning method, control positioning measurements and calculate the position of the target terminal using measurement results obtained from the target terminal itself and from dedicated *location measurement units*(LMU) in the BSs.

The LCS standards in 3GPP indicate a more formal view on the LCS functionality in different network elements and the corresponding interfaces between the network elements. For this reference model, see Figure 7.19.

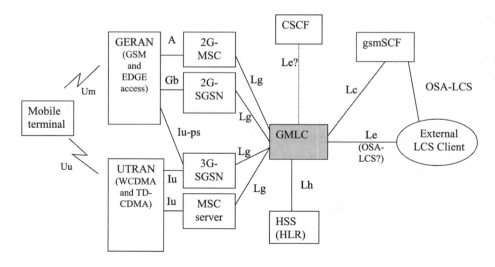

Figure 7.19. LCS reference model

LCS Functions in CN

The LCS functionality in CN is allocated to MSC/VLR (3GPP R99) and MSC server (3GPP R4), SGSN, HLR (3GPP R99), HSS (3GPP R4), and GMLC. The MSC/VLR and SGSN may initiate paging for LCS purposes and handle UE authorisation and call-related and non-call related position requests and charging.

The user can control if and how the position of his/her terminal is disclosed to different LCS clients. This is defined by LCS privacy classes in the subscription profile, which is kept in HLR or HSS. For value added services the user may want to be notified each time his/her position is being requested. The user may then choose to accept or decline a position request from an indicated LCS client. The network will send this privacy invocation request only to such terminals that are able to handle the request and can show it to the user. The terminal informs the network about its capability in this respect using UE classmark. The UE sends classmark 2 ("CN classmark") to the CN MSC/SGSN indicating whether the UE supports privacy verification or not (LCS value added capability).

UE also sends classmark 3 ("RAN classmark") to RNC (and to BSC + MSC in the case of GSM) indicating the supported LCS methods. This information is not needed in the CN. In the PS domain corresponding information elements are UE network capability and UE radio access capability.

MSC and SGSN handle charging and billing related to LCS. This includes charging and billing of both clients and subscribers including home subscribers and roaming subscribers from other PLMNs as well. MSC and SGSN also authorise the provision of positioning service for a particular UE in the same way as for other cellular services. The MSC/VLR handles LCS in the CN CS domain and SGSN supports LCS in the CN PS domain.

The HLR support for LCS is related to its conventional role, that is, it contains the subscription data and privacy profiles for the subscribers as well as information about which VLR and/or SGSN the mobile is currently registered. The GMLC requests routing information, i.e. in which VLR and/or SGSN the mobile is currently registered, from HLR via a Mobile Application Part (MAP) based interface.

The GMLC receives service requests from external LCS clients (or the mobile terminal itself) and forwards the positioning request to the MSC/VLR or SGSN as indicated by HLR. It has to be noted that the GMLC basically is a gateway element, which enables a connection of the network to any external client.

The GMLC identifies the LCS client within the UMTS PLMN by requesting client verification and authorisation from HLR, i.e. verifies that the LCS client is allowed to request positioning of the subscriber. GMLC determines if the position estimate received from the network satisfies the QoS requirement set by the LCS client. GMLC also provides flow control of positioning requests between simultaneous requests and it may convert the position information received from the network into local co-ordinates. GMLC may generate its own charging and billing related data for LCS.

LCS Functions in UTRAN

The serving RNC in UTRAN (or SMLC in GSM) selects the appropriate positioning method by which the indicated QoS level for the position request is met. The SRNC also controls how the positioning method is carried out in UTRAN and in UE.

When the OTDOA-IPDL method is used, the SRNC controls and defines the idle periods in different BSs to minimise the impact on UTRAN performance. This is done either according to pre-designed patterns or on-demand. RNC also co-ordinates the UTRAN resources involved in the positioning of the terminal.

In network based positioning methods the SRNC calculates the position estimate and indicates the achieved accuracy. It also controls a number of LMUs at BSs for the purpose of obtaining radio measurements needed to position or help to position a UE. Signalling between the SRNC and LMU is transferred via the Iub interface and in certain soft handover cases also via the Iur interface.

In UMTS, the LMU is normally integrated in BS and the stand-alone LMU will probably not be specified. The main functions of LMU is to measure the RTD between different BSs, the Absolute Time Differences (ATD) to a reference clock, or any other kind of radio interface timing measurement of the signals transmitted by BSs. Some measurement results returned by the LMU to SRNC can be used for more than one positioning method. All position and assistance measurements obtained by an LMU are sent only to the SRNC associated with the corresponding BS. SRNC controls the timing and any periodicity of the LMU measurements in BS, either directly or on a predefined basis.

LCS Functions in UE

Depending on the positioning method, the UE may be involved in the positioning procedures in various ways. For instance, in a network based positioning method, the UE does not calculate the position, whilst in a mobile based positioning method it does. The UE may also be equipped with a GPS receiver, which enables it to determine the position independently. The GPS functionality in UE may be enhanced by the network by sending GPS assistance information to the mobile. In one of the assisted GPS method alternatives the position of the UE having a GPS receiver is determined in the network based on the GPS measurement results reported by the UE.

The mobile terminal can request the network to determine its position (network assisted positioning) or the mobile itself may determine its own position by measuring signal timing and calculating its position using measurement results and assistance data from the network. The UE may also contain a LCS application, which acts as an LCS client.

The use of IPDL requires that the UE is capable of measuring and storing the radio signal timing during idle periods and to correlate different BCH codes with different idle period patterns. The UE needs to determine the arrival time of the first detectable path, both for the serving BS and other BSs that it detects. It must also be able to report the results back to the SRNC.

LCS Client and GMLC Interfaces

The LCS client provides services based on position information. The LCS client can be seen as a logical functional entity that requests position information from the cellular network (PLMN) for one or several target terminal(s). The request may contain a specified set of parameters defining the requested QoS and response time. The LCS client may reside in an entity (including the UE) within the PLMN or in an entity external to the PLMN.

In addition to the support for LCS by the main interfaces of UMTS, i.e. Uu, Iub, Iur and Iu, the following new LCS interfaces are defined for GMLC:

- The interface between the LCS client and GMLC is called Le and is shown in Figure 7.19. The LCS client sends the position request on this interface and GMLC sends the corresponding results back to the LCS client.
- Lh is a new interface between the GMLC and the HLR (HSS) as shown in Figure 7.19. The GMLC uses this interface to obtain the routing information data both for the CS and PS domain from the HLR.
- GMLC (in the home PLMN or in another PLMN) is connected to MSC/VLR over the Lg interface. The Lg interface is also used between GMLC and SGSN. GMLC forwards the position request for an indicated target terminal to the correct MSC/VLR or SGSN using the Lg interface and MSC/VLR and SGSN return the position result to GMLC.

Both Lh and Lg interfaces are based on the MAP protocol.

The Benefits of the LCS Architectural Solution in UMTS

The UMTS LCS system architecture is quite similar to the LCS architecture in GSM, but several optional network solutions were omitted in UMTS LCS in order to decrease the complexity of the solution. When compared to GSM LCS, the WCDMA-FDD based UTRAN brings both constraints and advantages for the cellular positioning applications, which are mainly related to the new radio access technology.

In UMTS, the SRNC has a key role in controlling the radio access, including radio resource management. The radio link measurements, including soft handover measurements, are terminated and controlled in SRNC. Since some 30–40% of all calls may be involved in soft handover the SRNC in many cases will have access to measurement results that can be used for positioning purposes. It was therefore seen reasonable to utilise this WCDMA-FDD advantage when considering the LCS architecture in UMTS. As a result, the functions corresponding to SMLC in GSM were allocated to SRNC in UMTS.

Also LCS assistance data may be utilised when optimising the system performance in association with radio resource management and this reinforces the placement of LCS functions in SRNC. The common architecture for LCS and radio resource management in RNC brings synergy benefits when optimising the usage of radio resources, signalling load and positioning mechanism and results in increased system performance.

The additional benefits of this architecture solution are the simplified signalling specifications on the Iu, Iur and Uu interfaces when compared with the many options in GSM. In UMTS the leading principle is to separate the radio related aspects from the CN aspects and this is valid also for LCS.

7.3.3. Service Capability – Mobile Application Execution Environment (MExE)

The terminals used in UMTS will not follow one standard; some of them will have the ability to present multimedia and some not. On the other hand, the services provide information in certain format(s). The question, which arises here, is how the information provided by the service can be presented in optimal format in the terminal. To solve this problem, the 3GPP specifications introduce Mobile Application Execution Environment (MExE).

The information MExE handles is MExE classmark, which is provided by the terminals supporting the MExE concept. There are two types of MExE classmark, classmark 1 and 2. The terminals supporting WAP type input and output provide MExE classmark 1: numeric keypad and limited screen size with fixed character amounts are the main indications in this respect. MExE classmark 2 is coming from the terminal if the terminal supports Java and thus utilises more flexible user interface alternatives: colour screen, full keypad and more powerful applications are the main indications in this respect.

From the MExE point of view, the "service" cloud in Figure 7.20 can be called MExE Service Environment (MSE). MSE does not have any exact configuration, the configuration depends on the services the service provider/carrier provider offer to its end users. In principle the MExE concept works with 2G networks too, but the UMTS network is the first step able to support the MSE more completely; within MSE and between the terminal and the MSE the information should be transferred by using HTTP.

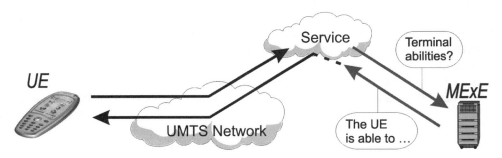

Figure 7.20. MExE principle

Another aspect related to MExE is environment enabling personalisation. Many Internet service providers offer the possibility to personalise, for instance, the portal starting page according to the user's needs. MExE does the same thing for the services, or to be exact, information presentation format.

The end user may adjust the terminal settings according to his/her needs and this information is stored locally at the USIM. The terminal type and its standard settings define the

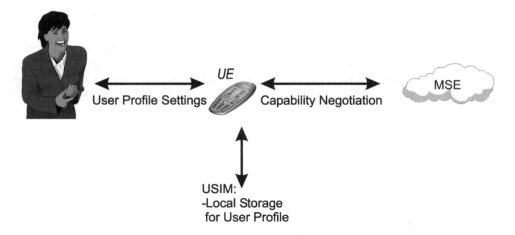

Figure 7.21. User profiles

terminal capabilities and MExE controls those on the network side. The final output the user receives is actually a result of the process whereby the requested information is put through two filters; firstly the terminal capabilities and secondly the user profile (Figure 7.21). For example, MExE classmark informs the network that the terminal is able to handle WAP (classmark 1). The user has defined *how* the received WAP information should be displayed (how many characters per line, how many lines, etc.). The MExE formats the information to be sent to the terminal according to these directives.

7.3.4. Service Capability – UMTS SIM Application Toolkit (USAT)

Physically the UMTS SIM (USIM) card is similar to the GSM SIM, but it has more advanced features. USIM contains more memory space, more processing power and it is downloadable. This creates room for new kinds of services not necessarily present in GSM networks. It should be noted that USIM is backward compatible with GSM SIM but the compatibility is not guaranteed in all conditions.

As shown in Figure 7.22 the USIM manipulation facility is implemented with separate service capability called USAT (UMTS SIM Application Toolkit). With this facility the

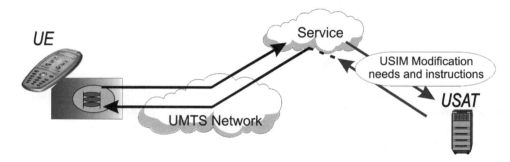

Figure 7.22. USAT principle

service provider may provide services, which, for example, modify the menu structures of the terminal. In this way the service provider may arrange easy "one-click-type" of WAP access, which in turn provides more services from the same service provider.

USAT is a very powerful tool and in that the service or carrier provider may completely change all of the subscription parameters such as IMSI number. This may be required in some services implemented with CAMEL but it is not recommended due to the high risks involved.

7.3.5. Service Capability – CAMEL SCE

CAMEL itself is a very large entity and is not a topic of this book as such. It however has remarkable role in UMTS networks and this is why some basic aspects of CAMEL should be highlighted in this context.

As was explained in Chapter 2 of this book, the cellular network evolution contains IN (Intelligent Network), which is mainly used for personalised services and their creation. IN implements this facility well but the IN services are carrier provider/service provider specific; IN services cannot be carried between networks. On the other hand, one of the key requirements of UMTS is automatic roaming with full service set. The concept adding this kind of mobility and service portability on top of IN is called CAMEL. If automatic roaming and Service Creation Environment (SCE) are used in UMTS networks every subscriber of the UMTS network has a CAMEL subscription, too.

Like IN, CAMEL takes care of circuit switched connection control. It has the same abilities as the IN and can perform as heavy or heavier involvement in that respect. Unlike IN, CAMEL is able to perform interworking with packet switched connections. This part of CAMEL use is not completely specified and remarkable changes are expected in this direction. CAMEL is able to interoperate with all the services used in GSM and SMS is one of those.

MM is an aspect which traditional IN does not support but CAMEL is able to transfer MM information within and between networks. CAMEL also interoperates with the UMTS network register functions, i.e. HLR and VLR thus being able to handle subscription data.

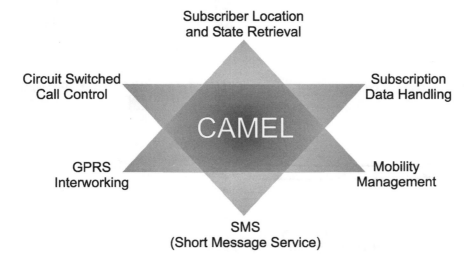

Figure 7.23.   CAMEL functioning environment

In addition to MM, CAMEL has direct interfaces with LCS, and thus it is able to utilise positioning information available through the LCS. The CAMEL interfaces with LCS are visible in Figure 7.23. The CAMEL interface in GSM is towards gsmSCF (Lc) and for UMTS towards CSCF (Le).

7.3.6. Example: Service Capabilities as Building Blocks

This example about the use of different service capabilities together is *highly hypothetical* but aims to introduce the way in which a single end-to-end service utilises service capabilities as building blocks within a 3GPP R4/R5 implementation.

Let us assume that a subscriber has a terminal able to receive and handle text, pictures, multimedia, i.e. it uses MExE classmark 2. The used subscription is pre-paid meaning that every single transaction debits a predefined account, no matter if the transaction was circuit or packet switched.

The service the subscriber uses here could have a conceptual name "local entertainment"; this service provides information about the restaurants, night-clubs, etc. within the area where the subscriber currently resides. Figure 7.24 describes the possible information flows.

The user has gained service "local entertainment" through, say, a service provider's web portal. This triggered the service provider to send the required piece of software to the USIM. In this context the USAT was used and, for instance, a menu item required to start "local entertainment" was added to the user profile in USIM. Now the user has all the means present in the UE to use the service when desired.

1. The user triggers the "local entertainment" service by sending a request concerning nearby located restaurants. This request is generated through a WAP-like browser facility within the terminal and thus the request is sent up to the WAP gateway. The WAP gateway

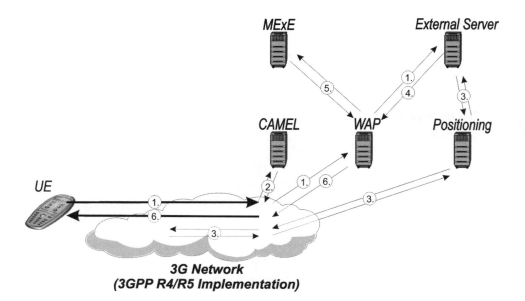

Figure 7.24.   Example about service capabilities

delivers the request further on to an external server containing information needed for the "local entertainment" service.

2. During the connection establishment the network triggers that this subscription is a pre-paid type and the subscriber's account is verified through CAMEL.

3. The external server needs the subscriber's positioning information in order to provide relevant information. It requests the position information from the GMLC, which in turn starts positioning procedures within the network. The GMLC performs the positioning of the user by using the methods described in Section 7.3.2. The GMLC returns the subscriber's positioning information to the external server. The external server determines with its application software (directory, database, etc.), which is the required information in this case.

4. When the information has been collected, the information is returned beck to the WAP gateway as a response to the subscriber's request.

5. The WAP gateway requests presentation instructions and subscriber profile information from the MExE. Hypothetically: if the terminal supports MExE classmark 1, the WAP gateway sends the information in textual format according to the terminal capabilities (display size, number of rows, number of characters per row, etc.). If the terminal supports MExE classmark 2 (personal Java) the WAP gateway may send information both in text and graphics and provide even more information, for example, restaurant logos, pictures, menus, etc.

6. When the terminal capabilities have been solved, the WAP gateway relays the information in desired format to the subscriber via the network. The terminal informs the subscriber, for instance, of the list of restaurants present in the current cell. The accuracy of the positioning depends on the positioning technology used (see Figure 7.15).

This description of one example service concludes our discussion about the UMTS network as a powerful platform for building value-added services to 3G users. Time will tell where the boundaries are for inventing new services based on the service capabilities provided by the UMTS network platform.

7.4. UMTS Services and Service Concepts

In this section we highlight some items related to the end-to-end services in UMTS. Though the UMTS network is a network for services, no one is in a position to tell what kind of services will be used through it during the years to come. In 3GPP R99 implementation the UMTS networks provide at least the same services as GSM. This means that from the end user point of view those services behave similarly but their implementation within the network is in many cases different than in GSM.

The evolution from GSM to UMTS, as described in Chapter 2, affects also the evolution of services. This development is, in principle, invisible to the end user but it causes remarkable changes in the network. If the evolution is studied against 3GPP specification releases it can be found that certain specification releases concentrate on certain parts of the network. The 3GPP R99 specifications concentrate on bringing the wideband radio access to the network and the CN is inherited from GSM with minor changes. The 3GPP R4 specifications instead concentrate on the CN and introduce new facilities like IP Multimedia Subsystem (IMS). In this context the handling of the services within the network will also change. For example, circuit switched voice calls could be delivered through IMS as VoIP calls. In 3GPP R5 there will again be changes to the access part of the network, but the CN will continue to develop further.

Table 7.4 presents the majority of GSM services and their treatment in 3GPP R99 implementation of UMTS.

Table 7.4 Services inherited from GSM and their treatment in 3GPP R99

Service(s)	Circuit Switched Domain R99	Packet Switched Domain R99
Speech (voice)	Supported	N/A (VoIP could be used in theory)
Facsimile	Supported	N/A (application level implementation like IP fax)
Alternate fax and speech	Supported	N/A
Circuit switched data bearer services	Supported	N/A
Supplementary services	Supported	N/A
USSD (unstructured supplementary service data)	Supported	N/A

7.4.1. Services and Conversational QoS Class

At least in the beginning of UMTS the most common service using conversational QoS class will still be the speech service. If speech service is implemented traditionally using the CN CS domain inherited from GSM, the QoS is not an issue as such. CS domain elements have already QoS "built into" the equipment. From the UMTS point of view, the situation is a bit more complicated because the speech coding algorithms used in GSM and UMTS are different. UMTS, however, emphasises the interoperability aspect between itself and GSM. Because the interoperability is implemented with inter-system handovers it leads to the situation where the speech coding method changes during the call. In order to maintain the quality of the connection the system may momentarily buffer the bit stream representing speech.

The way to solve this dilemma is to use an AMR (Adaptive Multi-Rate) coding technique as defined in the 3GPP specifications. The AMR codec is able to produce several source bit rates: 12.2, 10.2, 7.95, 7.40, 6.70, 5.90, 5.15 and 4.75 kb/s. From these rates, the 12.2 kb/s stream corresponds to GSM-EFR (Enhanced Full Rate) equal to the GSM EFR codec. Some of the other bit rates produced are also compatible with other 2G digital systems like US-TDMA and Japanese PDC. As far as speech service is concerned, the AMR speech codec can be considered like a "QoS converter" between different radio access technologies.

The next step is to convert the traditional speech service to a packet switched service, VoIP. The 3GPP R99 specifications assume that the speech service is still used traditionally, i.e. over circuit switched connections. The 3GPP R4/R5 specifications introduce completely different, universal CN structures favouring packet switching (please refer to Chapters 2 and 5). If the UMTS network is implemented according to 3GPP R4/R5, the speech service is carried like packet switched from the terminal through the network up to the Media Gateway (MGW) (Figure 7.25).

The MGW contains all the required and relevant mechanisms for different conversions; for instance, AMR is located there. Because surrounding networks are traditional circuit

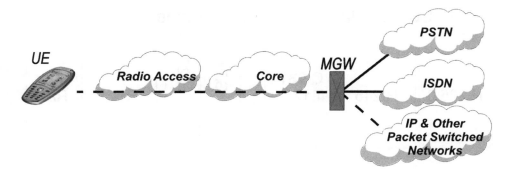

Figure 7.25.   VoIP and networks

switched like PSTN (Public Switched Telephony Network) and ISDN (Integrated Services Digital Network) the MGW also contains some conversion mechanisms for these, too. It should be noted that the conversion mechanisms are not only limited to the information flow, signalling between different networks must be converted, also. As far as VoIP is concerned, there are basically two alternatives for the call control signalling, H.323 specification and SIP (Session Initiation Protocol).

H.323 from ITU-T can be characterised like an "umbrella standard"; it contains all relevant standards required to define multimedia traffic delivery in such networks, which do not necessarily guarantee QoS.

H.323 defines four network elements, terminal, gateway, gatekeeper and Multipoint Control Unit (MCU). In UMTS networks these elements can be understood as functionalities and they are divided into network elements: H.323 terminal is an application located in the UE, H.323 gateway is partial functionality of the MGW, H.323 gatekeeper is equal to various load control functions located in the UTRAN and CN nodes and MCU is a separate functionality located in the CN.

The main function of the H.323 gateway is to arrange connections to other networks using different signalling scenarios, for instance, ISUP. The H.323 gatekeeper is responsible for controlling the resources of the network and their availability. In this respect the H.323 gatekeeper functionality is implemented in UTRAN admission control, for instance.

The H.323 works in the protocol environment described in the Figure 7.26. H.323 is carried over the IP stack. TCP is used in the context of signalling because it is more reliable, though slower. The UDP layer is used for user data; errors are allowed but speed is more essential. On top of this, the H.323 covers various standards for various purposes.

Control plane:

- H.225.0: this standard handles session management, data multiplexing and connection quality observation aspects.
- H.245: this forms a control channel used to deliver information about supported audio and video coding formats, channel management and flow control of the connection. H.245 is very often implemented with limited facilities.
- Q.931, T.120: H.323 gatekeeper may implement various standards for connection establishment. The mandatory one is Q.931 defining ISDN-type of call set-up. The others like T.120 for data connections are not mandatory.

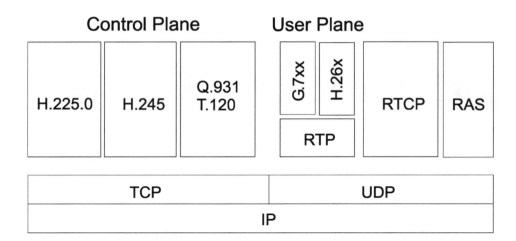

Figure 7.26.   H.323 protocol environment

The H.323 user plane is meant for audio and video transfer. To transfer these, the H.323 defines so called Real-time Transport Protocol (RTP), which defines frame structures and synchronisation issues. The RTP is used by different codecs, most of them are not mandatory. The only mandatory codec is G.711, which is required to establish traditional 64 or 56 kb/s PCM coding for voice connections. The rest of the protocols, RTCP (Real-time Control Protocol) and RAS (Registration, Admission and Status) are used for connection quality control, bandwidth control and status information transfer.

As can be seen, the H.323 implementation of VoIP and other real-time services over packet switched connection may be relatively complicated. Another alternative when considering especially VoIP is Session Initiation Protocol (SIP). SIP is an "Internet world counterpart" for H.323; it aims to fulfil the same tasks as H.323 but has been implemented from an IP basis. The advantage of SIP when compared to H.323 is that SIP is simpler; connection establishment does not require as many messages (= time) as in H.323. SIP can be considered in the context of IMS but the specifications in 3GPP concerning this are being processed.

Figure 7.27 introduces a simplified sample of how the MGW treats SIP and ISUP in the context of VoIP. The SIP session starts with an INVITE message, which corresponds to ISUP Initial Address Message (IAM). The ISUP Address Complete Message (ACM) is interpreted as a positive acknowledgement on the SIP side and so on. The main idea is that every SIP message has one counterpart on the ISUP side and vice versa. If H.323 is used in similar conversion, the H.323 side may use even tens of messages to generate one ISUP message towards the other network. This kind of delay could be critical in certain situations.

In this respect the SIP is a better alternative for VoIP and possibly for other real-time multimedia sessions, too. On the other hand, H.323 (because it is an older standard) has better product support and terminal availability.

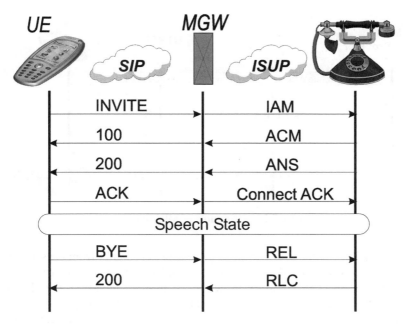

Figure 7.27.   Signalling conversion sample SIP ⇔ ISUP

7.4.2. Services and Streaming QoS Class

If a service uses conversational QoS class it does not transfer any files from one end to another literally. The first QoS class handling file transfer through the connection is the streaming class. Because complete file download takes time and thus causes delay, there must be mechanisms to open and handle files when they are not completely transferred from the source to the destination. This is what the streaming QoS class covers. Typical services or applications using streaming class are those handling big files but showing/playing/handling a limited part of it in time. Also the services offering a multicast (one sender and many receivers simultaneously) type of service utilise streaming class if delay is not an issue. To minimise the possible delay effects the streaming class services are mostly unidirectional; delay exists but it does not cause any harm because the interactivity is missing.

Figure 7.28 shows a rough example how an MP3 music file can be listened to by using streaming QoS class. The application located in the UE sends the user's request concerning a certain MP3 file to the network through which the request finally arrives at the desired service provider. As an acknowledgement, the service provider starts downloading the desired MP3 file. When using streaming QoS class the whole MP3 file is not required in the UE at the same time. The point here is that the packet format data arrives fluently enough to the UE and the packets are in the correct order. The data stream is buffered in two places for two different purposes: in UTRAN the data is buffered in order to control the radio interface access and its load and in the UE the streaming class application constructs buffers from the data stream in order to present the data to the user "real-time-like", i.e. continuously without any cuts and jitter. When the application in the UE has buffered enough data to start the playback, the data is presented/played to the user the through selected application. At the same time the stream-

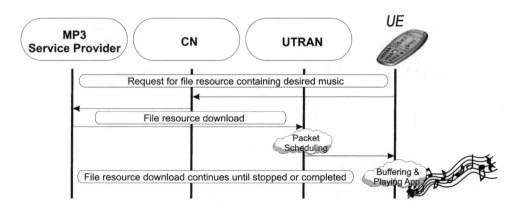

Figure 7.28.   Example of MP3 resource use

ing class application continues to download more data to its buffer from the service provider. The data download is continued until the file is completed or the user interrupts the process.

Streaming class services could, in principle, be implemented with the H.323 protocol stack shown in Figure 7.26, but the implementation would be complicated. On the other hand, streaming class services have developed a lot in the Internet environment. The Internet offers very limited bandwidth for standard users and different streaming techniques have been developed to eliminate this limitation. The streaming techniques used in the Internet environment are based on the "client–server" thinking; the end user has an application able to handle streaming bit flow, to store it and handle it later on when required. The "client–server" part of this kind of streaming is that when the user likes to listen/view/handle the bit stream, the application starts data buffering and opens its interface to the end user. In other words, in Internet-type-of-streaming, the real streaming is done by the application located within the user terminal, in H.323 based streaming the streaming is done within the network and the end user has just the interface to handle the streaming bit flow with predefined way.

Both of these types will be in use because different services using streaming QoS class will set their own requirements. On the other hand, the aim is to utilise the Internet-type-of-streaming as much as possible because it makes network implementation simpler and thus more cost effective.

7.4.3. Services and Interactive/Background Class

The interactive class can afford quite remarkable and variable delays. The services utilising this QoS class are traditional request-response-type of services like web surfing (request: URL address of a desired web page, response: the contents of the web page), WAP services (logic is the same as before) and other services requiring opening of server connections. A new set of services called location based services utilise this QoS class. Location based service mechanisms require relatively heavy involvement by the UMTS network and these procedures are described in more detail in Section 7.3.2.

The background QoS class provides the very basic connectivity level, where delay and Bit Error Rate (BER) are, in practise, meaningless if the network is in a position to provide error correction functionality. The services utilising this QoS class are typically various procedures including file downloads. For instance, e-mail receiving is one of these.

7.4.4. Service Concept: Virtual Home Environment (VHE)

At first glance, the idea of VHE sounds simple: the user must always have the same, personalised individual interface through the terminal no matter through which (UMTS) network he/she is attached to.

Figure 7.29 shows the rough principle of VHE. Though this service concept is very simple from the subscriber point of view, it sets high requirements for the networks involved.

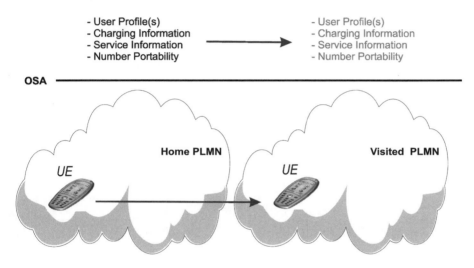

Figure 7.29. The principle of VHE

In principle, VHE is a concept taking care that, for a certain subscriber, the subscriber profile(s), charging information, service information and numbering information is either transferred or is transparently available between networks. All these listed items are complicated matters when handling between networks. The common denominator and key to handle them between the networks is CAMEL and actually the VHE sets the requirement that CAMEL should be furnished in every UMTS network implementing VHE. As shown in Figures 7.7 and 7.8, the service capabilities are accessible through CAMEL. Thus:

* Subscriber profile(s) are handled in MSE
* Location based services information can be transferred between networks through CAMEL
* Other service information (provisioned services and their usage information) is transferred between the networks through CAMEL
* Number portability can be handled, in principle, through CAMEL

As can be seen, the VHE is difficult to implement and from those above-mentioned items the number portability is most probably the most difficult issue due to the different numbering schemes used throughout the world.

8

Security in the UMTS Environment

Security has always been closely associated with mobile networks due to the very nature of wireless communications, which are easily accessible by not only the intended mobile phone users but also by any potential eavesdropper. However, security is a much wider set of issues, which need to be considered by all the players in the mobile communication business. In this chapter we will consider those both from the carrier/service providers' and from the end users' point of view. At the end of this chapter lawful interception is also discussed, which is yet another aspect to mobile communications, which have become part of the modern society.

The analogue mobile networks (first generation (1G)) did not have too many security related features and in those times the subscribers did not necessarily understand the meaning of secure connections. On the other hand, radio path (which as such is the most unsecured part of the system) can be secured very effectively but this arrangement has its pros and cons. If the radio connection is point-to-point type (between two users) a dedicated security arrangement can be used. This is the way the military radio connections are most often secured. In public cellular networks the radio path is always point-to-multipoint, i.e. the same transmitted signal is received by many terminals and thus any dedicated security system on the radio path is not usable as such (Figure 8.1).

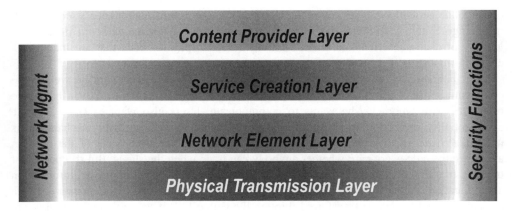

Figure 8.1. Security and its role in 3G

In basic GSM, security is concentrated on the radio path security, i.e. the access network part. In UMTS networks, security is a larger topic. The access network connection must naturally be secured but in addition to this, security must be taken into account in many other respects, too. The commercial model/business chain explained later in this chapter leads to the situation where sometimes very sensitive information is transferred between different parties and networks. From a security point of view, this can be considered as a severe security risk. Also local and international authorities are setting their directives in this respect. UMTS continues to integrate together telecommunication and data communication worlds and this could also create threats against security. In IP world security has been an issue for many years and numerous security threats have been identified and defence mechanisms have been developed.

The rest of this chapter is organised as follows: in the first section we study the access security of UMTS. This includes both secure user access to UMTS networks and security of the connection at the access network level. The security of UMTS access is based on the GSM access security model but several enhancements have been made.

The second section deals with security at the network element layer. Here the main issue is how to secure connections inside a UMTS network and between networks that are controlled by different mobile operators. As the UMTS Core Network (CN) (on its 3GPP R99 level) is based on the GSM CN, there are no major security enhancements on this front when compared to the second generation (2G) case. The situation changes a lot in subsequent releases of UMTS as the core network architecture evolves. In this book we only give a preliminary view of the mechanisms that will be used.

The third section provides an overview of the security issues and mechanisms at the upper layers of service and content providers. These are largely independent of the structure of the UMTS network itself but, nevertheless, they play a major role in the overall security of the system. Also in the third section, we briefly discuss access security of IP multimedia subsystems inside the UMTS network. Finally, in the fourth section we show how the regulatory requirement for lawful interception capability is satisfied in UMTS.

8.1. Access Security in UMTS

The radio access technology will change from TDMA to WCDMA when the third generation (3G) mobile networks are introduced. Despite this shift, requirements for access *security* will not change. Still in UMTS, it is required that end users of the system are *authenticated*, i.e. the identity of each subscriber is verified; nobody wants to pay for calls that are made by a cheating impostor.

The *confidentiality* of voice calls is protected in the radio access network, as well as the confidentiality of transmitted user data. This means that the user has control over the choice of parties with whom he/she wants to communicate. Users also want to *know* that the confidentiality protection is really applied: *visibility* of applied security mechanisms is needed. *Privacy* of the user's whereabouts is generally appreciated. Most of the time an average citizen does not care whether anybody can trace where he/she is. But if a persistent tracking of users were to occur, he/she would be quite irritated. Similarly, exact information about location of people would be useful, e.g. for burglars. Also, privacy of the user data is a critical issue when data is transferred through the network. Privacy and confidentiality are largely synonymous in this presentation.

Availability of the UMTS access is clearly important for a subscriber who is paying for it. Network operators consider *reliability* of the network functionality to be important: they want control inside the network to function effectively. This is guaranteed by the *integrity* of all radio network signalling; it is checked that all control messages have been created by authorised elements of the network. In general, integrity checking protects against any manipulation of a message, e.g. insertion, deletion or substitution.

The most important ingredient in providing security for network operators and subscribers is *cryptography*. That consists of various techniques which all have roots in the science and art of *secret writing*. It is sometimes useful to make communication deliberately incomprehensive, i.e. to use ciphering (or, synonymously, encryption). This is the most effective way to protect communications against malicious purposes.

8.1.1. Legacy from 2G

The transition from analogue 1G systems to digital 2G systems made it possible, among many other things, to utilise advanced cryptographic methods. The most important security features in the GSM system are:

- Authentication of the user
- Encryption of communication in the radio interface
- Use of temporary identities

As GSM and other 2G systems became more and more successful, the usefulness of these basic security features also became more and more evident. Naturally, it has been a leading principle in specification work of UMTS security to carry these features over to the new system.

The success of GSM also emphasised finally the limitations of its security. A popular technology is also tempting for fraudsters. The properties of GSM that have been most criticised on the security front are the following:

- Active attacks towards the network are possible in principle: this refers to somebody who has the required equipment to masquerade as a legitimate network element and/or legitimate user terminal (see Figure 8.2 for an example scenario).
- Sensitive control data, e.g. *keys* used for radio interface ciphering, are sent between different networks without ciphering.
- Some parts of the security architecture are kept secret, e.g. the cryptographic algorithms: this does not create trust in them in the long run because they are not publicly available for

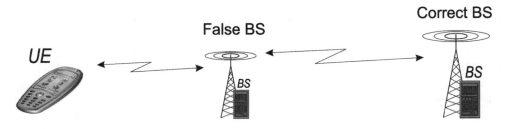

Figure 8.2.   Active attack

analysis by novel methods and, on the other hand, global secrets tend to be revealed sooner or later.

- Keys used for radio interface ciphering become eventually vulnerable to massive *brute force* attacks where somebody tries all the possible keys until one matches.

These limitations were left in the GSM system on purpose. The threat imposed because of them was estimated inferior in comparison to the added cost of trying to circumvent them. However, as the technology advances, the attackers gain access to better tools. That is why the outcome of a similar comparison between cost and security led to a different conclusion in the 3G case.

In UMTS, countermeasures for perceived weaknesses in GSM are developed. This is another leading principle that has guided the design of the 3G security architecture. The most important security features in the access security of UMTS are the following:

- Mutual authentication of the user and the network
- Use of temporary identities
- Radio access network encryption
- Protection of signalling integrity inside UTRAN

Note that publicly available cryptographic algorithms are used for encryption and integrity protection. Algorithms for mutual authentication are operator-specific.

Each of these features are described in the Sections 8.1.2, 8.1.4, 8.1.5 and 8.1.6. Details can be found in the 3GPP specification TS 33.102.

8.1.2. Mutual Authentication

There are three entities involved in the authentication mechanism of the UMTS system:

- Home network
- Serving network (SN)
- Terminal, more specifically USIM (typically in a smart card)

The basic idea is that the SN checks the subscriber's identity (as in GSM) by a so-called challenge-and-response technique while the terminal checks that SN has been authorised by the home network to do so. The latter part is a new feature in UMTS (compared to GSM) and through it the terminal can check that it is connected to a legitimate network.

The mutual authentication protocol itself does not prevent the scenario in Figure 8.2 but it (in combination with the other security mechanisms) guarantees that the active attacker cannot get any real benefit out of the situation. The only possible gain for the attacker is to be able to disturb the connection but clearly no protocol methods exist that can circumvent this type of attack completely. For instance, an attacker can implement a malicious action of this kind by radio jamming.

The cornerstone of the authentication mechanism is a *master key K* that is shared between the USIM of the user and the home network database. This is a permanent secret with a length of 128 bits. The key K is never transferred out from the two locations. For instance, the user has no knowledge of her/his master key.

At the same time with mutual authentication, keys for encryption and integrity checking are derived. These are temporary keys with the same length of 128 bits. New keys are derived from the permanent key K during every authentication event. It is a basic principle in cryptography to limit the use of a permanent key to a minimum and instead derive temporary keys from it for protection of bulk data.

We describe now the Authentication and Key Agreement (AKA) mechanism at a general level. The authentication procedure can be started after the user is identified in the serving network. The identification occurs when the identity of the user, i.e. permanent identity IMSI or temporary identity TMSI, has been transmitted to VLR or SGSN. Then VLR or SGSN sends an *authentication data request* to the Authentication Centre (AuC) in the home network.

The AuC contains master keys of users and based on the knowledge of IMSI the AuC is able to generate *authentication vectors* for the user. The generation process contains executions of several cryptographic algorithms, which are described later in more detail. The generated vectors are sent back to VLR/SGSN in the *authentication data response*. This process is depicted in Figure 8.3. These control messages are carried on the MAP protocol.

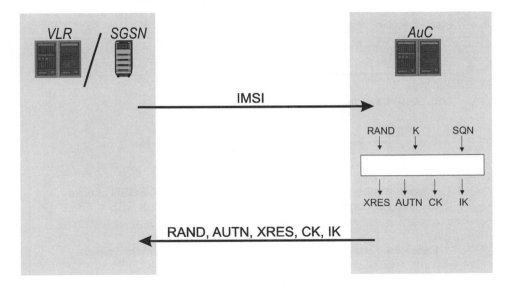

Figure 8.3. Authentication data request and authentication data response

In the serving network, one authentication vector is needed for each authentication instance, i.e. for each run of the authentication procedure. This means the (potentially long distance) signalling between SN and the AuC is not needed for every authentication event and it can in principle be done independently of the user actions after the initial registration. Indeed, the VLR/SGSN may fetch new authentication vectors from AuC well before the number of stored vectors runs out.

The serving network (VLR or SGSN) sends a *user authentication request* to the terminal. This message contains two parameters from the authentication vector, called RAND and AUTN. These parameters are transferred into the USIM that exists inside a tamper-resistant environment, i.e. in the UMTS IC card (UICC). The USIM contains the master key K, and using it with the parameters RAND and AUTN as inputs, USIM carries out a computation that resembles the generation of authentication vectors in AuC. This process also contains executions of several algorithms, as is the case in the corresponding AuC computation. As the result of the computation USIM is able to verify whether the parameter AUTN was indeed generated in AuC and, in the positive case, the computed parameter RES is sent back to VLR/SGSN in the *user authentication response*. Now the VLR/SGSN is able to compare user response RES with the expected response XRES which is part of the authentication vector. In the case of match, authentication ends positively. This part of the process is depicted in Figure 8.4.

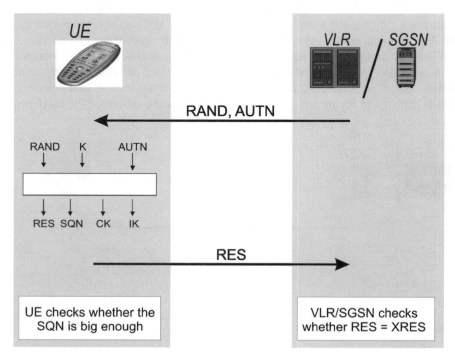

Figure 8.4.   User authentication request and user authentication response

The keys for radio access network encryption and integrity protection, namely CK and IK, are created as a by-product in the authentication process. These temporary keys are included in the authentication vector and, thus, are transferred to the VLR/SGSN. These keys are later transferred further into the RNC in the radio access network when the encryption and integrity protection are started. On the other side, the USIM is able to compute CK and IK as well after it has obtained RAND (and verified it through AUTN). Temporary keys are subsequently transferred from USIM to the mobile equipment where the encryption and integrity protection algorithms are implemented.

8.1.3. Cryptography for Authentication

Next we take a closer look at the generation of authentication vectors in the AuC. A more detailed illustration of the process is given in Figure 8.5. The process begins by picking up a correct sequence number SQN. Roughly speaking, what is required is that

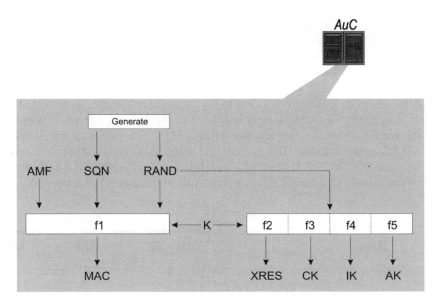

Figure 8.5.   Authentication vector generation

the sequence numbers are picked up in an increasing order. The purpose of the sequence number is to prove later to the user that the generated authentication vector is *fresh*, i.e. it has not been used before. In parallel with the choice of the sequence number, a random bit string RAND with length of 128 is generated. This is a demanding task by itself but in this presentation we just assume that a cryptographic pseudorandom generator is in use which produces large amounts of unpredictable output bits when a good physical random source is available as a seed for it.

The key concept in the authentication vector computation is a *one-way function*. This is a mathematical function, which is relatively easy to compute but practically impossible to invert. In other words, given the input parameters there exists a fast algorithm to compute the output parameters but, on the other hand, if the output is known there exist no efficient algorithms to deduce any input that would produce the output. Of course, one simple algorithm to find the correct input is to try all possible choices until one gives the wanted output. Clearly this *exhaustive search* algorithm becomes extremely inefficient as the length of the input increases.

In total, five one-way functions are used to compute the authentication vector. These functions are denoted f1, f2, f3, f4 and f5. The function f1 differs from the other four in the number of input parameters. It takes four input parameters: master key K, random number RAND, sequence number SQN and finally an administrative *authentication management field* AMF. All other functions from f2 to f5 take only K and RAND as inputs. The requirement of the one-way property is common to all functions f1–f5 and all of them can be built around the same *core* function. However, it is essential that they differ from each other in a fundamental way: from the output of one function no information about the outputs of the other functions can be deduced. The output of f1 is Message Authentication Code (MAC, 64 bits), outputs of f2, f3, f4 and f5 are, respectively, XRES (32–128 bits), CK (128 bits), IK (128 bits) and AK (64 bits). The authentication vector consists of the parameters RAND, XRES, CK, IK and AUTN. The last one is obtained by concatenating three different parameters: SQN added bit-by-bit to AK, AMF and MAC.

Now we take a closer look into the handling of the authentication on the USIM side. This is illustrated in Figure 8.6. The same functions f1–f5 are involved also on this side but in a slightly different order. The function f5 has to be computed before the function f1 since f5 is used to conceal SQN. This concealment is needed in order to prevent eavesdroppers from getting information about the user identity through SQN. The output of the function f1 is marked XMAC on the user side. This is compared to the MAC received from the network as part of the parameter AUTN. If there is a match it implies RAND and AUTN have been created by some entity that knows *K*, i.e. by the AuC of the user's home network.

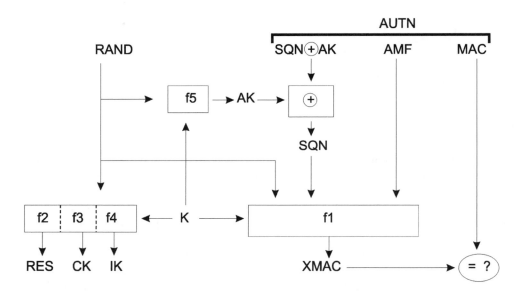

Figure 8.6. Authentication handling in USIM

Still there is a possibility that some attacker who has recorded an earlier authentication event replays the pair of RAND and AUTN. As mentioned above, the sequence number protects against this threat. The USIM should simply check that it has not seen the same SQN before. The easiest way to check this is to require that the sequence numbers appear in an increasing order. It is also possible that the USIM allows some SQNs to arrive out-of-order if it maintains e.g. a short list of greatest sequence numbers received so far. Since the transfer of authentication vectors from the AuC and the actual use of these vectors for authentication are done somewhat independently there are several reasons why it is possible that authentication vectors are used in a different order than that in which they are originally generated. The most obvious reason for such a case is a consequence of the fact that mobility management functions for CS and PS domains are independent of each other. This implies that authentication vectors are fetched to VLR and SGSN independently of each other and are also used independently.

The choice of the algorithms f1–f5 is in principle operator-specific. This is because they are used only in the AuC and in the USIM and the same home operator controls them both. An example set of algorithms (called MILENAGE) exists in the 3GPP specification TS 35.206. Sequence number management is also operator-specific in principle. There are two basic strategies in creating sequence numbers: each user may have an individual sequence number or sequence number generation may be based on a global counter, e.g. universal time. A combination of these two strategies is also possible: for instance, the most significant part of the SQN is user-specific but the least significant part is based on a global counter.

The mutual authentication mechanism is based on two parameters stored both in AuC and USIM: a static master key K and a dynamic sequence number SQN. It is vital that these parameters are maintained in a synchronised manner on both sides. For the static K this is easy but it is possible that the dynamic information about sequence numbers runs out of synchronisation for whatever reason. As a consequence, the authentication would fail. A specific *re-synchronisation procedure* is used in this case (see Figure 8.7). By using the master key K as the basis for secure communication, the USIM informs the AuC of its current SQN.

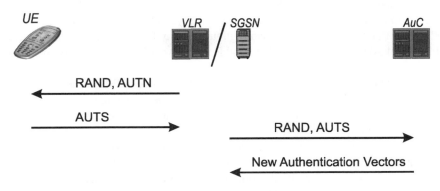

Figure 8.7.   Re-synchronisation procedure

A parameter AUTS is delivered during the re-synchronisation. It contains two parts: the sequence number of the USIM concealed by AK and a message authentication code MAC-S computed by another one-way function f1* from the input parameters SQN, K, RAND and AMF. The last two parameters are obtained from the failed authentication event. The one-way function f1* has to be different from f1 because otherwise recorded AUTN-parameters could in principle be accepted as valid AUTS-parameters in the re-synchronisation and an attacker could at least disturb the authentication mechanism.

8.1.4. Temporary Identities

The permanent identity of the user in UMTS is IMSI as is the case also for GSM. However, the identification of the user in UTRAN is in almost all cases done by temporary identities, TMSI in the CS domain or P-TMSI in the PS domain. This implies that confidentiality of the user identity is protected almost always against passive eavesdroppers. Initial registration is an exceptional case where a temporary identity cannot be used since the network does not yet know the permanent identity of the user. After that it is in principle possible to use temporary identities.

The mechanism works as follows. Assume the user has been identified in the serving network by IMSI already. Then the serving network (VLR or SGSN) allocates a temporary identity (TMSI or P-TMSI) for the user and maintains the association between the permanent identity and the temporary identity. The latter is only significant locally and each VLR or SGSN simply takes care that it does not allocate the same TMSI/P-TMSI to two different users simultaneously. The allocated temporary identity is transferred to the user once the encryption is turned on. This identity is then used in both uplink and downlink signalling until a new TMSI (or P-TMSI) is allocated by the network.

The allocation of a new temporary identity is acknowledged by the terminal and after that the old temporary identity is removed from the VLR (or SGSN). If allocation acknowledgement is not received by VLR/SGSN it shall keep both the old and new (P-)TMSIs and accept either of them in uplink signalling. In downlink signalling, IMSI must be used because the network does not know which temporary identity is currently stored in the terminal. In this case, VLR/SGSN tells the terminal to delete any stored TMSI/P-TMSI and a new re-allocation follows.

Still one problem remains: how does the serving network obtain the IMSI in the first place. Since the temporary identity has a meaning only locally, the identity of the local area has to be appended to it in order to obtain an unambiguous identity for the user. This means Location Area Identity (LAI) is appended to TMSI and Routing Area Identity (RAI) is appended to P-TMSI.

If the UE arrives into a new area then the association between IMSI and (P-)TMSI can be fetched from the old location or routing area if its address is known to the new area (based on LAI or RAI). If the address is not known or a connection to the old area cannot be established, then IMSI must be requested from the UE.

There are some specific places, e.g. airports where lots of IMSIs may be transmitted over the radio interface as people are switching on their mobile phones after the flight. On the other hand, tracking of people is usually also otherwise easier in those places: for instance, watch who is walking out of which plane! Altogether, the user identity confidentiality mechanism in UMTS does not give 100% protection but it offers a relatively good protection level. Note that the protection against an active attacker is not very good since the attacker may pretend to be a new serving network to which the user has to reveal his/her permanent identity. The mutual authentication mechanism does not help here since the user has to be identified before he/she can be authenticated.

8.1.5. UTRAN Encryption

Once the user and the network have authenticated each other they may begin secure communication. As described earlier, a cipher key CK is shared between the core network and the terminal after a successful authentication event. Before encryption can begin, the communicating parties have to agree on the encryption algorithm also. Fortunately, in UMTS implemented according to 3GPP R99, only one algorithm is defined. The encryption/decryption takes place in the terminal and in the RNC on the network side. This means that the cipher key CK has to be transferred from CN to the radio access network. This is done in a specific RANAP message called *security mode command*. After the RNC has obtained CK it can switch on the encryption by sending an RRC *security mode command* to the terminal.

The UMTS encryption mechanism is based on a *stream cipher* concept as described in

Figure 8.8. This means the plaintext data is added bit-by-bit to random-looking *mask* data, which are generated based on the cipher key CK and a few other parameters. This type of encryption has the advantage that the mask data can be generated even before the actual plaintext is known. Then the final encryption is a very fast bit operation. The decryption on the receiving side is done in exactly the same way since adding the mask bits twice has the same result as adding zeros.

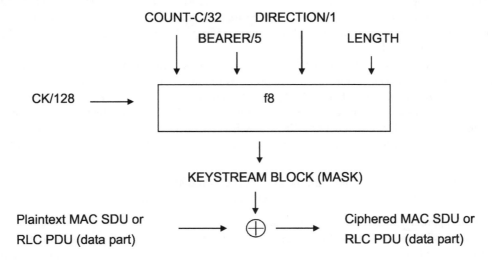

Figure 8.8.   Stream cipher in UMTS

As the mask data does not depend on the plaintext at all there has to be another input parameter, which changes every time a new mask is generated. Otherwise, two different plaintexts, say P_1 and P_2, would be protected by the same mask. Then the following unwanted phenomenon would happen: if we add P_1 to P_2 bit-by-bit, and we do the same to their encrypted counterparts, then the result bit string is exactly the same in both cases. This is, again, due to the fact that two identical masks cancel each other in the bit-by-bit addition. Therefore, the bit-by-bit sum of P_1 and P_2 would become known to any attacker who eavesdrops on the corresponding encrypted messages on the radio interface. Typically, if two bit strings of meaningful data are added to each other bit-by-bit, both of them can be totally revealed from the result bit string. Hence, this would imply a break of the encryption for the two messages P_1 and P_2.

The encryption occurs in either the *medium access control* layer (MAC) or in the *radio link control* layer (RLC). In both cases, there is a counter that changes for each PDU. In MAC this is *connection frame number* (CFN) and in RLC a specific *RLC sequence number* (RLC-SN). If these counters are used as such as input for mask generation the problem explained in the previous paragraph would still occur since these counters wrap around quickly. This is why a longer counter called a *hyperframe number* (HFN) is introduced. It is incremented whenever the short counter (CFN in MAC and RLC-SN in RLC) wraps around. The combination of HFN and the shorter counter is called COUNT-C and it is used as an ever-changing input to the mask generation inside the encryption mechanism.

In principle, the longer counter HFN could also eventually wrap around. Fortunately, it is reset to zero whenever a new key is generated during the authentication and key agreement procedure. The authentication events are frequent enough to rule out the possibility of HFN wrap-around.

The radio bearer identity BEARER is also needed as an input to the encryption algorithm since the counters for different radio bearers are maintained independently of each other. If the input BEARER would not be in use, then this would again lead to a situation where the same set of input parameters would be fed into the algorithm, and the same mask would be produced more than once. Consequently, the problem explained above would occur and the messages (this time in different radio bearers) encrypted with the same mask would be exposed to the attacker.

The core of the encryption mechanism is the mask generation algorithm which is denoted as the function f8. The specification is publicly available as 3GPP TS 35.201 and it is based on a novel *block cipher* called KASUMI (for which there is another 3GPP specification TS 35.202). This block cipher transforms a 64-bit input to a 64-bit output. The transformation is controlled by the 128-bit cipher key CK. If CK is not known there are no efficient algorithms to compute the output from the input or vice versa. In principle, the transformation can be done if either:

- All possible keys are tried until the correct one is found; or
- An enormous table of all 2^{64} input–output pairs is collected in somehow

Both approaches are impossible in practice.

It is possible that authentication is not performed in the beginning of the connection. In this case the previous cipher key is used for encryption. The key is stored in the USIM between the connections. Also, the most significant part of the greatest HFN used so far is stored in the USIM. For the next connection, the stored value is incremented by one and used as the starting value for the most significant part of HFN.

A ciphering indicator is used in terminals to show user whether encryption is applied or not, thus providing some visibility of the security mechanisms. Note that although use of ciphering is highly recommended it is still optional for the UMTS network.

8.1.6. Integrity Protection of RRC Signalling

The purpose of the integrity protection is to authenticate individual control messages. This is important since a separate authentication procedure gives assurance of the identities of the communicating parties only at the time of the authentication. Figure 8.2 can be used to illustrate the issue: a *man-in-the-middle* (false BS) acts as a simple relay and delivers all messages in their correct form until the authentication procedure is completely executed. After that, the man-in-the-middle may begin to manipulate messages freely. However, if messages are protected individually, deliberate manipulation of messages can be observed and false messages can be discarded.

The integrity protection is implemented at the RRC layer. Thus, it is used between the terminal and RNC, just like encryption. The integrity key IK is generated during the authentication and key agreement procedure, again similar to the cipher key. Also, IK is transferred to the RNC together with CK in security mode command.

The integrity protection mechanism is based on the concept of a *message authentication code*. This is a one-way function, which is controlled by the secret key IK. The function is denoted by f9 and its output is MAC-I: a 32-bit random-looking bit string. The MAC-I is appended to each RRC message and it is also generated and checked on the receiving side. Any change in the input parameters influence the MAC-I in an unpredictable way. The function f9 is depicted in Figure 8.9. Its inputs are IK, the RRC message itself, a counter COUNT-I, direction bit (uplink/downlink) and a random number FRESH. The parameter COUNT-I resembles the corresponding counter for encryption. Its most significant part is a HFN and the four least significant bits consist of the RRC sequence number. Altogether, COUNT-I protects against replay of earlier control messages: it guarantees that the set of values for input parameters is different for each run of the integrity protection function f9.

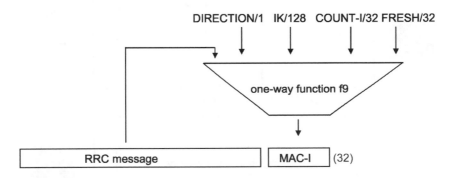

DIRECTION/1   IK/128   COUNT-I/32 FRESH/32

one-way function f9

RRC message   MAC-I   (32)

Figure 8.9.   Message authentication code

The parameter FRESH is chosen by the RNC and transmitted to the UE. It is needed to protect the network against a maliciously chosen start value for COUNT-I. Indeed, the most significant part of HFN is stored in the USIM between connections. An attacker could masquerade as the USIM and send a false value to the network forcing the starting value of HFN to be too small. If the authentication procedure is not run, then the old IK is used. This would create a chance for the attacker to replay RRC signalling messages from earlier connections with recorded MAC-I values if only the parameter FRESH was not present. By choosing FRESH randomly, the RNC is protected against these kinds of replay attacks, which are based on recording of earlier connections. As already explained, the ever-increasing counter COUNT-I protects against replay attacks based on recording during the same connection as FRESH stays constant over a single connection. Note that the radio bearer identity is not used as an input parameter for the integrity algorithm, although it is an input parameter for the encryption algorithm. Because there are several parallel radio bearers for the control plane also, this seems to leave room for a possible replay of control messages that were recorded within the same RRC connection but on a different radio bearer. However, this is not the case, since the radio bearer identity is always appended to the message when the message authentication code is calculated (although it is not transmitted with the message). Therefore, the radio bearer identity has an effect on the MAC-I value and we have protection also against replay attacks based on recordings on different radio bearers.

It is clear that there are a few RRC control messages whose integrity cannot be protected by the mechanism. Indeed, messages sent before the integrity key IK is in place cannot be protected. A typical example is the *RRC connection request* message.

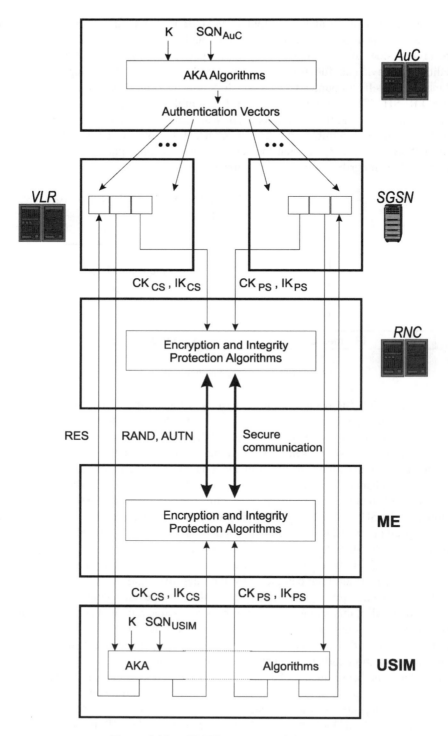

Figure 8.10.   UMTS access security summary

The algorithm for integrity protection is based on the same core function as the encryption. Indeed, the KASUMI block cipher is used in a special mode to create a message authentication code function.

An integrity protection mechanism in UTRAN is not applied for the user plane. This is due to performance reasons. On the other hand, there is a specific (integrity protected) control plane procedure, which is used for *periodic local authentication*. As a result of the procedure, the amount of data sent during the RRC connection is checked. Hence, the *volume* of transmitted user data is integrity protected.

8.1.7. Summary of Access Security

We conclude this section by presenting a schematic overview of the most important access security mechanisms and their relationships to each other. For the sake of clarity, many parameters are not shown in Figure 8.10. For instance, HFN and FRESH are important parameters, which are transmitted between different elements and yet they are omitted from the figure.

8.2. Security Aspects on the System and Network Level

In this section we briefly discuss potential security threats on the network level and how to protect against them. The target is to provide confidentiality and integrity protection for communication between different network elements. These elements belong either to the same network or to two different networks. Especially in the latter case, fully standardised security solutions are needed to ensure interoperability.

The 3G business chain recognises several parties involved in the business, "subscriber", "carrier provider", "service provider" and "content provider" (Figure 8.11). If we think of a service the subscriber uses and is charged for, every party in this chain is involved; the carrier provider provides the platform through which the connection is established. The service provider arranges USIM (identification information) and actual service. Generally speaking a service requires content and this is taken care of by the content provider. The carrier, service

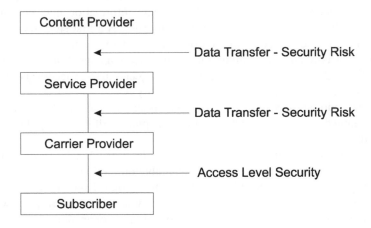

Figure 8.11.   Security risks in the business chain

and content provider could physically be the same company but this is not necessarily the case. In practice, all kinds of combinations concerning these three parties exist in the GSM world. It is clear that the subscriber is charged for the service and its use, but the other parties in the business chain share the income. The system controlling the monetary flow between the carrier, service and control provider is a potential security risk, since very sensitive information is transferred between these parties.

8.2.1. Typical Security Attacks

Many threats towards the communication between UMTS network elements are similar to the threats against communication on the application layer. Clearly there are also big differences, already, between various applications, but the attack examples presented in this subsection have to be taken into account in all cases.

There are plenty of ways people try to perform security attacks; only imagination and effective protection are the limits. The following list presents some examples of these. Please note that the list is definitely not exhaustive.

- Social engineering
- Electronic eavesdropping (sniffing)
- Spoofing
- Session hijacking
- Denial of service (DoS)

Social engineering is not usually identified as a security risk by the normal user/subscriber, though it has an essential role in many attacks. As far as subscribers are concerned, social engineering means, for instance, the ways to gain access to one's terminal, i.e. to get the PIN number one way or another. The subscriber is able to protect his/herself against social engineering by having a PIN enquiry in the terminal active and keeping the PIN in safe place separated from the terminal. Social engineering is also surprisingly common in the network side; sometimes the people working with network elements in the operator's premises will receive weird calls where the caller explains he needs a user ID and password to an equipment there and the person responsible is on holiday or otherwise unreachable. Often these calls are social engineering where secure information may end up in wrong hands. Actually, this is the way the hacker typically tries as a first step. As a result, social engineering may give some access to vital or non-vital network elements. In the IP world social engineering is relatively common but in the telecommunications world social engineering is not so widely used; the equipment is not subject to public access and the personnel maintaining them are very aware of their responsibility.

Electronic eavesdropping, also known as sniffing, is another commonly used attack method as already explained in previous sections of this chapter. Sniffing is very difficult to detect and prevent physically (Figure 8.12).

With sniffing the hacker aims to collect, for example, user ID and password information. Unfortunately, sniffing programs are publicly available on the Internet for anyone to download. A sniffing program itself is only a tool and in the right hands it is used for network monitoring and possible fault detection. In the wrong hands it is a powerful tool with which a hacker is able to silently monitor great amounts of Internet connections.

The information gathered by sniffing can be utilised in the next step with a hacking method

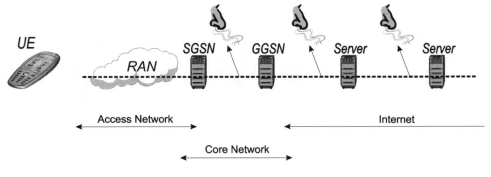

Figure 8.12.   Sniffing

called spoofing. Spoofing as a method means that a hacker uses someone else's IP address and, *receives* packets from the other users. In other words, the hacker replaces the correct receiver in the connection.

Surely the hacker could try to open a connection with someone else's IP address but it is more difficult and very often this kind of "one-way traffic" described in Figure 8.13 is also useful for hacking purposes. Nowadays when people do a lot of work remotely at home, this could be one way to gain access to company information when the employer and employee exchange data over the Internet.

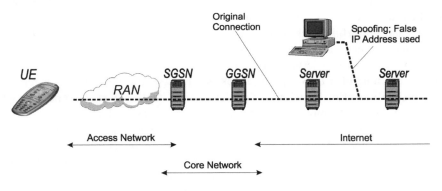

Figure 8.13.   Spoofing

If the attacker takes one step further from spoofing he is trying session hijacking where he attempts to take over an existing connection. As mentioned earlier, even a strong authentication mechanism in the beginning of the connection does not protect against hijacking it later. We also need integrity protection over the whole session.

In the Denial of Service (DoS) attack the hacker does not aim to collect information, rather he is aiming to cause harm and inconvenience to other users and service provider(s) (Figure 8.14).

In a typical DoS attack the hacker generates disturbing traffic, which, in the worst case, jams the target server in such a way that it is not able to provide a service any more. The idea behind this is, for instance, to fill the server's service request queue with requests and then ignore all acknowledgements the server sends back. Consequently, the server occupies

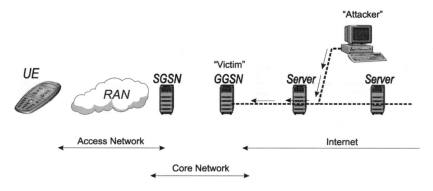

Figure 8.14. Denial of Service (DoS)

resources for the incoming connection, which never occurs. When timers of the connection expire, the resources are freed to serve another connection attempt. When the buffer containing connection attempts are continuously filled with new requests, the server is actually stuck with these requests and it is not able to provide a "real" service.

There are also more sophisticated DoS attacks and plenty of tools are available for DoS attacks on the Internet. Altogether, protection against DoS is very difficult.

An advanced DoS attack may be combined together with the other methods described above. For example, DoS is initiated from "stolen" IP addresses and in a distributed way where there are even hundreds of computers attending to the DoS attack. DoS is a very dangerous and powerful attack and it easily causes large economical losses. In February 2000 two large Internet companies, experienced serious DoS attacks and their connections were not available for real service during that time.

Another example of a DoS type of attack is targeting towards Internet search engines, which are simply overloaded with inexact data queries that are just sent one after another. This is a very rude and simple way to get search engine servers stuck.

The preceding text described briefly the most common security attacks one may experience. These are threats, which cannot be ignored in any case since the tools to implement these are free-of-charge and available on the Internet. What is there to do then? Security should be thought of as a chain, where the entire system (communication security, data security and signalling security) is only as strong as the weakest link. In other words, everything must be secure; this covers algorithms, protocols, links, end-to-end paths, applications, etc.

8.2.2. Overview of 3GPP Release 5 Network Domain Security

It was mentioned earlier that one of the weaknesses in GSM security architecture stems from the fact that authentication data is transmitted unprotected between different networks. For instance, cipher keys are used to protect the traffic on the radio interface but these keys are themselves transmitted in clear between networks. The reason for this state of affairs lies in the closed nature of SS7 networks: only a relatively small number of large institutions have access to them. In UMTS Release 99 the core network structure is still very much like that in GSM and that is why no major enhancements were done in the security of traffic between CNs.

In the subsequent releases of UMTS the situation changes: core network structure evolves and IP becomes the dominant protocol on the network layer. Although this does not mean that signalling between different core networks would be carried over truly open connections there is certainly a shift towards easier access to core network traffic. There are much more players involved and there exists a community of hackers who are skillful in IP issues.

The basic tool in protection of network domain traffic is the IPSEC protocol. It provides confidentiality and integrity of communication in the IP layer. Also, communicating parties can authenticate each other using IPSEC. The critical issue is key management: how to generate, exchange and distribute various keys needed in algorithms that are used to provide confidentiality and integrity protection. We present a brief introduction to IPSEC in the next section.

In addition to the protection of IP-based networks, security of purely SS7-based networks is enhanced in UMTS Release 5. In particular, a specific security mechanism has been developed for the MAP protocol. This is called MAPSEC and it provides confidentiality and integrity protection. The key management to support MAPSEC is planned to be provided by similar techniques that are also used to support IPSEC.

8.2.3. IPSEC

IP security is standardised by IETF (Internet Engineering Task Force). It consists of a dozen RFCs and it is a mandatory part of IPv6. In IPv4 IPSEC can be used as an optional "add-on" mechanism to provide security in the IP layer. The main IPSEC components are the following:

- Authentication Header (AH)
- Encapsulation Security Payload (ESP)
- Internet Key Exchange (IKE)

The purpose of IPSEC is to protect IP packets: this is done by ESP and/or AH. In short, ESP provides both confidentiality and integrity protection while AH provides only the latter. There are more fine-grained differences between the two but clearly ESP and AH are largely overlapping mechanisms. One of the reasons for including this kind of redundancy in IPSEC standards is export control: there have been severe restrictions on the export of confidentiality protection mechanisms in most countries while integrity protection mechanisms have typically been free from restrictions. Currently such export restrictions are largely dropped and as a consequence, the importance of AH compared to ESP is decreasing.

Both ESP and AH need keys. More generally, a notion of a Security Association (SA) is essential in IPSEC. In addition to the encryption and authentication keys, SA contains information about the used algorithm, lifetime of the keys and the SA itself, a sequence number to protect against replay attacks, etc.

Security associations must be negotiated before ESP or AH can be used, one for each direction of communication. This is done in a secure way by IKE protocol. There are several modes of IKE but the idea is clear: the communicating parties are able to generate "working keys" and SAs, which are used in protection of subsequent communication. IKE is based on the ingenious idea of public key cryptography where secret keys for secure communication can be exchanged over an insecure channel. However, authentication of the parties who run IKE cannot be done without some long-term keys. These are typically based on either manual

Transport Mode:

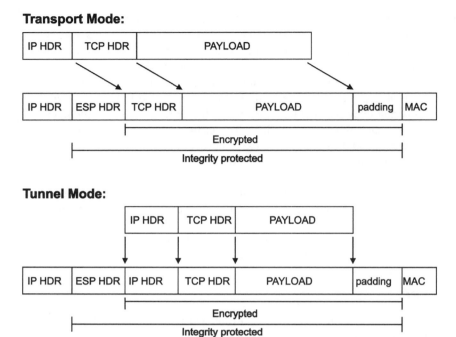

Figure 8.15.   Encapsulation Security Payload (ESP)

exchange of a shared secret or Public Key Infrastructure (PKI) and certificates. Both solutions are clearly non-trivial: the first requires lots of configuration effort while the latter implies dedicated infrastructure elements with special functionality.

The negotiation of SAs by IKE is independent of the purpose for which these SAs are used. This is why IKE can be used also for negotiation of keys and SAs to be used in, e.g. MAPSEC.

We conclude this section by describing ESP a bit more. There are two ESP modes, transport mode and tunnel mode. The transport mode functions basically as follows. Everything in an IP packet except the IP header is encrypted. Then a new ESP header is added between the IP header and the encrypted part: this contains, e.g. the information about the SA in use. Also, the encryption typically adds some bits onto the end of the packet. Finally, a Message Authentication Code (MAC) is calculated over everything except the IP header and it is appended to the end of the packet. In the receiving end, the integrity is checked first. This is done by removing the IP header from the beginning of the packet and the MAC from the end of the packet, then running the MAC function (based on the information in the ESP header) over the rest and comparing the result to the MAC in the packet. If the outcome of the integrity check is positive, then the ESP header is removed and the rest is decrypted (again based on information in the ESP header). See Figure 8.15 for an illustration.

The tunnel mode differs from the transport mode in the following way. A new IP header is added onto the beginning of the packet. Then the same operations as in the transport mode are carried out for the new packet. This means that the IP header of the original packet is protected, as illustrated in Figure 8.15.

The transport mode is the basic use case of ESP between two end points. However, when applied in the UMTS networks there are two problems: the communicating network elements have to:

- Know the IP address of each other;
- Implement all the IPSEC functionality.

The typical use case of the tunnel mode is related to the concept of a Virtual Private Network (VPN). IPSEC is used between two middle nodes (security gateways) and the end-to-end protection is provided implicitly as the whole end-to-end packet is inside the payload of the packet that is protected between the gateways. Clearly, in this scenario each element has to trust the corresponding gateway. Also, the leg from the end-point element to the gateway has to be protected by other means, e.g. it is implemented in a physically protected environment. The preferred protection method for UMTS core network control messages is to use ESP in the tunnel mode between security gateways.

8.2.4. MAPSEC

Now we briefly discuss some basic properties of MAPSEC. The idea is to protect the confidentiality and integrity of MAP operations. In protection mode 2 of MAPSEC, both confidentiality and integrity are protected while in protection mode 1, only integrity is protected. In protection mode 0 there is no protection.

For confidentiality, the payload of original MAP operation is encrypted. A security header is added to indicate how decryption should be done and, in general to point to the correct MAPSEC security association. For integrity, a MAC is calculated over the cleartext payload of the original MAP operation and the security header. A time variant parameter is used to protect against replay attacks.

The security associations for MAPSEC are created using the IKE protocol. This is done with dedicated key management entities called Key Administration Centres (KAC), which negotiate keys on behalf of all other CN elements in the same network. This calls for some enhancements inside IKE, which were not yet finalised in IETF in early 2001.

8.3. Protection of Applications and Services

The security in the network is implemented according to the OSI (Open System Interconnection) model, i.e. each OSI layer has its own means for security purposes (Figure 8.16). To simplify this, there are two main branches, which are link-by-link security and end-to-end security.

In link-by-link protection the idea is that a connection (path) is formed by communication links. Everything going through a particular link is protected by, e.g. encryption, and if this is done for every link along a path, the whole connection is secured. The drawback here is that if traffic is misrouted the contents of the link can be exposed and this is not necessarily visible to the end-user.

If security is applied in the higher OSI layers, we talk about end-to-end security. In this case the data remains protected, for instance, it is encrypted until it reaches its destination. The transmitted data is well secured but now it is possible to carry out *traffic analysis*: one is able to find out, for instance, who sent it, who received it, when the

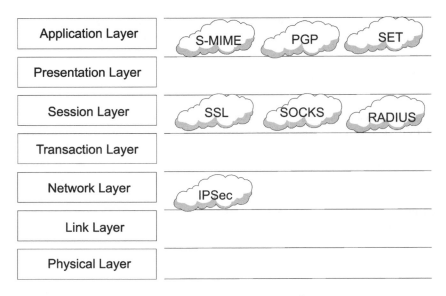

Figure 8.16.   Security protocols in different OSI layers (examples)

transmission was done and how often the transmission has been done. If a total network security solution is implemented, one must combine both link-to-link and end-to-end security mechanisms. Encryption of each link makes any analysis of the routing information impossible, while end-to-end encryption reduces the threat of unencrypted data at the various nodes in the network.

In certain cases, such as cellular network, link layer authentication is required in order to control access to the cellular network. In the 3G network, the access level security procedures take care of link level encryption and integrity protection within the access network. Within the CN and in the connections to the other network various methods can be used including those discussed in the previous section.

8.3.1. Application Layer Security

At the application layer most common security mechanisms include S-MIME, PGP, PEM, S-HTTP and SET. The S-MIME (Secured Multipurpose Internet Mail Extension) is a security protocol that adds digital signatures and encryption to Internet MIME messages. It was originally developed by RSA Data Security Inc. and it is based on triple-DES encryption (triple Data Encryption Standard) and X.509 digital certificates. S-MIME uses an RSA public-key encryption method and Diffie–Hellman system for key management. SHA-1 (Secure Hash Algorithm #1) is adopted for data integrity protection purposes.

Version two of S-MIME, S-MIME v2 has been widely adopted in the Internet mail industry. The current work on S-MIME is being done in the IETF S-MIME Working Group. S-MIME v3 was standardised in July 1999.

PGP (Pretty Good Privacy) is a freeware for e-mail security originally designed by Mr. Philip Zimmermann. It provides privacy and authentication to the e-mail by using encryption and digital signatures. PGP uses IDEA (International Data Encryption Algorithm) for encryption and RSA for key management and digital signatures. Data integrity is protected based on, e.g. MD5 algorithm (Message Digest #5).

One of the most interesting aspects of PGP is its distributed approach to key management. There are no key certificate authorities (CA); instead every user generates and distributes his/ her own public key. Users sign each other's public keys, creating an interconnected community of PGP users. The benefit of this mechanism is that there is no CA that everyone has to trust. Each user keeps a collection of signed public keys in a file called a public-key ring. Each key in the ring has a key legitimacy field that indicates the degree to which the particular user trusts the validity of the key. The user sets this field manually. The weakest link of the PGP is key revocation. If someone's private key is compromised (e.g. stolen), a key revocation certificate has to be sent out. Unfortunately it is not at all guaranteed that everyone, who may use the public key that corresponds to the compromised private key, receives the key revocation in time.

So far, the end-to-end security protocols briefly introduced have concentrated on e-mail. The Internet, especially in the context of mobile networks, is seen as a very promising area for electronic payments. Monetary transactions require very secure connections and common trust. For this purpose Visa and MasterCard have together with their technology partners developed a protocol named SET (Secure Electronic Transactions). SET has already been adopted by major banks and financial bodies world-wide as an Internet e-commerce standard. SET uses digital certificates to validate the identity of all parties in a SET transaction. The original SET protocol seems complex from the end user's point of view and a new version has been recently developed which is easier to use.

8.3.2. Security for Session Layer

At the session layer of an OSI stack one could use, e.g. SSL (Secure Socket Layer) or SOCKS (Socket Security). The SSL was originally developed by Netscape Communications Corporation to provide privacy and reliability between two communicating applications at the Internet session layer. SSL uses public-key encryption to exchange a session key between the client and server and this session key is used to encrypt the http (hypertext transfer protocol) transaction. Each transaction uses a different session key so, that if someone manages to decrypt a transaction the session is still secure; only one transaction is exposed. In the past the encryption has used 40-bit (rest of the world) or 128-bit (US) secret keys but the situation changes as the export restrictions are loosened.

A new Internet standard called TLS (Transport Layer Security) has been developed from SSL. There is also a specific modified version called WTLS (Wireless Transport Layer Security), which is optimised for the wireless environment. It is used to secure WAP sessions.

SOCKS is a protocol that enables hosts on one side of a SOCKS proxy server to gain full access to hosts on the other side of the SOCKS server without requiring direct IP reachability.

8.3.3. AAA Security Mechanisms

RADIUS (Remote Authentication Dial-In User Service) is described in RFC 2138 as a protocol for carrying authentication, authorisation and configuration information between NASs (Network Access Server) and shared authentication servers (Figure 8.17). RADIUS provides a method that allows multiple dial-in NAS devices to share a common authentication database.

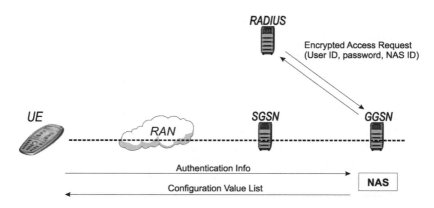

Figure 8.17.   RADIUS protocol data flow

It was originally developed by Livingston Incorporated for their Portmaster line of NAS products and has been widely deployed in many vendors' products over recent years.

Client–server model is used for a NAS to manage user connections. A NAS operates as a client of RADIUS, which is responsible for passing user information to designated RADIUS servers. RADIUS servers are responsible for receiving user connection requests, authenticating the user, and then returning all configuration information (type of service, SLIP, PPP, Login User-ID, values to deliver the service) necessary for the client to deliver service to the user.

The IETT AAA working group (Authorisation, Authentication and Accounting) has developed a new protocol called DIAMETER, which should gradually replace RADIUS as the dominant AAA mechanism.

8.3.4. IMS Security

A major addition in UMTS Release 5 is the introduction of an IP Multimedia Subsystem (IMS). It is designed to be independent of the access network technology. Therefore, security for IMS cannot be provided solely by UMTS Release 99 security features.

The control for IMS is carried by Session Initiation Protocol (SIP). From the UMTS Release 99 point of view, SIP control for IMS is user plane traffic. This implies that, in the case where UTRAN is used to access IMS, confidentiality of IMS signalling is protected inside UTRAN but integrity protection is not in use. In some other access networks there may be no protection of IMS signalling in the access network level.

Mutual authentication is the cornerstone security feature for IMS. Also, confidentiality and integrity of SIP signalling are protected in UMTS release 5 and keys agreed during the mutual authentication procedure are used to support these features. Early in 2001, the preferred authentication and key agreement mechanism for IMS is planned to be the same one that is used for bearer level authentication in UMTS release 99. Although the mechanism is the same, execution of authentication in IMS is independent of the release 99 authentication procedure run between SGSN and USIM. Furthermore, different master keys may be used. Whether confidentiality and/or integrity protection of SIP messages is provided in SIP itself or by some lower layer has not yet been decided.

8.4. Lawful Interception

One crucial goal in security is to encrypt information in such a way that it is available only for the correct receiver. So far, in this chapter we have presented mechanisms, which show how this is done in 3G networks.

On the other hand, there are many countries where local authorities and laws set limits for encryption, thus actually limiting the security level the network is able to offer. In addition/ parallel to this, the local regulations may set a requirement that the authorities must have a way for access to sensitive information and subscriber observation, i.e. the authorities must have a chance to listen to the calls or monitor data traffic, both circuit and packet switched. In GSM, this kind of arrangement was added on top of the existing system afterwards but in 3G this localised requirement has been taken into account from the very beginning.

As a whole the lawful interception arrangement consists of three parts, the interception equipment/functionality, mediation part and the interception information (Figure 8.18). The interception equipment/functionality collects items indicated in Tables 8.1 and 8.2. The local requirements define what is needed and what is not. This filtering is taken care of by local mediation devices which show then only the information that the local regulations define. This filtered information is called interception information.

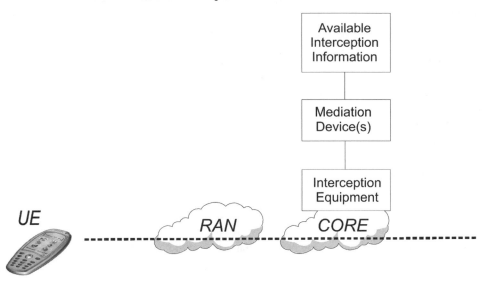

Figure 8.18.   Lawful interception arrangement – rough principle

In addition to the items listed in Tables 8.1 and 8.2 there are some other regulations related to the position of a subscriber. For example, in some countries it will in the future be mandatory for the system to show the position of a subscriber in case of an emergency call with an accuracy of 50 m. This kind of accuracy is not achieved with normal mobility management (MM) methods and instead the 3G network contains equipment specialised for this purpose. It is called a positioning system. In this book, the most common positioning methods are presented in Chapter 4 and positioning system architecture is discussed in Chapter 7.

Table 8.1   Items collected for circuit switched transaction lawful interception

Collected item	Explanation
Observed MSISDN	Target identifier with the MSISDN of the target subscriber (monitored subscriber).
Observed IMSI	Target identifier with the IMSI of the target subscriber (monitored subscriber).
Observed IMEI	Target identifier with the IMEI of the target subscriber (monitored subscriber), It shall be checked for each call over the radio interface
Event type	Description which type of event is delivered: establishment, answer, supplementary service, handover, release, SMS, location update, Subscriber Controlled Input (SCI)
Event date	Date of the event generation in the 3G MSC
Event time	Time of the event generation in the 3G MSC
Dialled number	Dialled phone number before digit modification, IN-modification, etc.
Connected number	Number of the answering party
Other party address	Directory number of the other party for MOC calling party for MTC
Call direction	Information whether the monitored subscriber is calling or called, e.g. MOC/MTC or originating/terminating party
Correlation number	
Location information	The service area identity and/or location area identity that is present at the 3G MSC at the time of event record production
Basic service	Information about teleservice or bearer service
Supplementary service	Supplementary services used by the target, e.g. CF, CW, ECT
Forwarded to number	Forwarded to number at CF
Call release reason	Call release reason of the target call
SMS message	The SMS content with header which is sent with the SMS-service
Redirecting number	The number which invokes the call forwarding towards the target. This is provided if available
SCI	Non-call related SCI which the 3G MSC receives from the ME

Table 8.2   Items collected for packet switched transaction lawful interception

Collected item	Explanation
Observed MSISDN	MSISDN of the target subscriber (monitored subscriber)
Observed IMSI	IMSI of the target subscriber (monitored subscriber)
Observed IMEI	IMEI of the target subscriber (monitored subscriber), it shall be checked for each activation over the radio interface
Event type	Description which type of event is delivered: PDP attach, PDP detach, PDP context activation. Start of intercept with PDP context active, PDP context deactivation, SMS, cell and/or RA update
Event date	Date of the event generation in the 3G SGSN and/or GGSN
Event time	Time of the event generation in the 3G SGSN and/or GGSN

Table 8.2 (*continued*)

Collected item	Explanation
PDP address	The PDP address of the target subscriber. Note that this address might be dynamic
Access Point Name (APN)	The APN of the access point (typically the GGSN of the other party)
Routing area code	The routing area code of the target defines the RA in a PLMN
PDP type	The used PDP type. For instance, PPP, IP or X.25
Correlation number	
SMS	The SMS content with header which is sent with the SMS-service. The header also includes the SMS-centre address
Failed attach reason	Reason for failed attach of the target subscriber
Failed context activation reason	Reason for failed context activation of the target subscriber
IAs	The observed interception areas

9

UMTS Protocols

In previous chapters the UMTS network has been discussed in the subnetwork and network element centric manner. The functional split between the network elements has been presented by allocation of major system-level management functions to the network elements. At the same time the interfaces between the different kind of network elements have already been introduced.

In this chapter we now take an interface-centric view of the UMTS network by focusing on the system protocols. The UMTS protocols are used to control the execution of network functions in a co-ordinated manner across the system interfaces.

9.1. Protocol Reference Architecture in 3GPP R99

Since the protocol specification work was divided between the various groups in the 3GPP technical organisation it was quite natural that each of the groups developed protocol reference architecture for those protocols for which it has got the mandate to do the specification work. This resulted in three major areas with protocol reference models of their own.

Before presenting a combined model for all UMTS network protocols, which will be followed throughout this chapter, let's have a look at each of those three reference models. It appears that among them some major protocol architectural concepts have been introduced and we can benefit from those when deriving the combined protocol architecture.

9.1.1. The Radio Interface Protocol Reference Model

The UTRAN radio interface is based on the WCDMA radio technology, which has some fundamental differences as compared to any 2G radio access technology. The key aspect to be controlled by the radio interface protocols is multiplexing of traffic flows of different kinds and different origins. To ensure effective control of multiplexing the layering of duties has been applied, thus resulting in the three-layer protocol reference model illustrated in Figure 9.1.

The layers are named simply according to their position in the architecture as radio interface layer 1, layer 2 and layer 3. Although the well-known OSI principle of layering has been applied in this design, these three layers cannot as such be considered as the three lowest layers of the OSI model. The layers have well-defined responsibilities and interfaces with each other, which makes it possible to name them as follows:

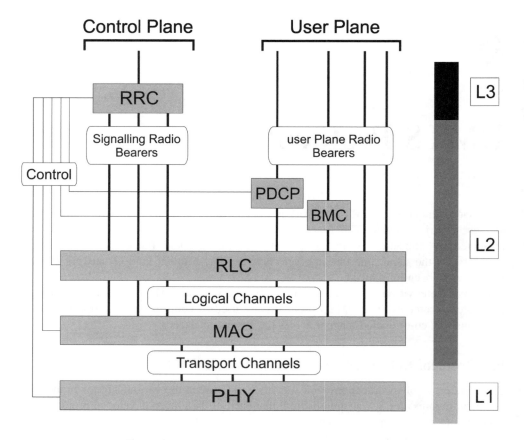

Figure 9.1.   Radio interface protocol reference model

- L1 – radio physical layer
- L2 – radio link layer
- L3 – radio network layer

As shown in Figure 9.1 the physical layer provides its services as a set of WCDMA transport channels. This makes the physical layer responsible for the first multiplexing function: to map the flows from transport channels to WCDMA physical channels and vice versa. The mapping of transport channels onto physical channels was explained in Chapter 4.

The radio link layer is another multiplexing layer and it does make a major contribution to dynamic sharing of the capacity in the WCDMA radio interface. Instead of the wide variety of L1 transport channels this layer allows the upper layer to see only a set of radio bearers, along which different kinds of traffic can be easily transmitted over the radio. This UTRAN radio bearer is part of the UMTS system-wide bearer architecture, which was explained already in Chapter 1.

The Medium Access Control (MAC) sublayer controls the use of transport block capacity by ensuring that the capacity allocation decisions (done on the UTRAN side) are executed promptly on both sides of the radio interface. The Radio Link Control (RLC) sublayer then adds regular link layer functions onto the logical channels provided by the MAC sublayer. Due to the characteristics of the radio transmission some special ingredients have been added to the RLC sublayer functionality.

For L3 control protocol (signalling) purposes the RLC service is adequate as such, but for domain-specific user data additional convergence protocols may be needed to accomplish the full radio bearer service. For CS domain data (e.g. transcoded speech) the convergence function is null, but for the PS domain an additional convergence sublayer is needed. This Packet Data Convergence Protocol (PDCP) sublayer makes the UMTS radio interface applicable to carry Internet Protocol (IP) data packets. Another convergence protocol (BMC) has been specified for message broadcast and multicast domains. The scheduling and delivery of cell broadcast messages to UEs is the main task of this protocol.

The separation of control signalling from user data is another key design criteria applied for a long time in protocol engineering. In this way the protocols used for control purposes become part of the system-wide control plane whereas protocols carrying the end-user data belong to the user plane.

The L3 control plane protocol the is Radio Resource Control (RRC) protocol. As shown in Figure 9.1 an RRC protocol entity on both the UE and UTRAN side has control interfaces with all other protocol entities. Whenever the protocol entity to be controlled by RRC is located in another UTRAN network element, there is a need to support this control mechanism with standardized protocols. In other cases the control interfaces are internal to a single UE or UTRAN element and hence not precisely standardised, but their existence is crucial to allow the RRC sublayer to carry out its task as executor of the radio resource management decisions.

9.1.2. UTRAN Protocol Reference Model

The access network, UTRAN, has the overall responsibility of WCDMA radio resources. UTRAN carries out the radio resource management and control among its own network elements and creates the radio access bearers to allow communication between UEs and the Core Network (CN) across the whole UTRAN infrastructure. This communication is structured according to the generic UTRAN protocol reference model illustrated in Figure 9.2. The model is generic in the sense, that the architectural elements – though not the specific protocols – are the same for all the UTRAN interfaces: Iu, Iub and Iur. The specific protocols are referred to in Figure 9.2 only by using their 3GPP specification numbers, which also indicates the idea of sharing the common protocol architecture.

The main elements of the UTRAN protocol architecture are layers and planes. The general layering principle is applied in the UTRAN design first of all to distinguish between two major parts of the protocol stack instead of going down into individual layers. Therefore, all the lower layer protocols together compose what is called transport network layer of UTRAN. The name transport is used here pretty much in its OSI sense, i.e. to cover the set of protocols allowing non-adjacent network nodes to communicate across an internetwork composed of possibly many different kind of subnetworks. Another, more pragmatic design aspect has been to include into the transport network layer all those protocols, which were selected from existing protocol suites instead of having to design them specially for UMTS purposes.

The other set of protocols on top of the transport network layer is – consistently with the radio interface protocol model – called radio network layer. These protocols have been

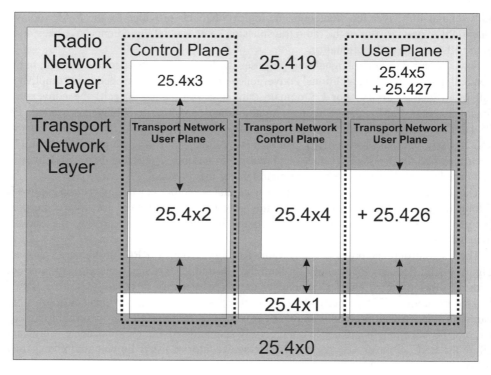

Figure 9.2. Generic UTRAN protocol reference model. Legend to the 3GPP specifications: $x = 1$ Iu protocol; $x = 2$ Iur protocols; $x = 3$ Iub protocols

carefully designed for the UMTS system and the common task for them is to control the management and use of radio access bearers across the various UTRAN interfaces.

Figure 9.2 also distinguishes between control plane and user plane protocols. This design aspect, which was already discussed before on radio interface protocols, has been applied over all UTRAN interfaces. Within the transport network layer the protocols have been selected keeping in mind the various properties required to support both control and user plane protocols in the most appropriate way. For control plane protocols this means reliability as a selection criteria and for user plane protocols, on the other hand, Quality of Service (QoS) support from the transport network has been targeted in protocol selection. The control protocols for the radio network layer has been specially designed to meet the radio and bearer control requirements. The control protocols for the UTRAN radio network layer are commonly named UTRAN Application Part (AP) protocols. On the other hand, the user plane protocols within the radio network layer, which share the common property of being responsible of efficient transfer of user data frames, are commonly knows as UTRAN frame protocols

The UTRAN AP and frame protocols together form the UMTS access stratum, which covers all those communication aspects, which are dependent on the selected radio access technology. Within the generic UMTS radio access bearers the non-access stratum protocols are then used for direct transfer of signalling and transparent flow of user data frames between UEs and the CN. This is achieved by the encapsulation of higher layer payload into UTRAN protocol messages.

The UTRAN control plane protocols have been designed to follow the client–server principle. Regarding the Iu interface UTRAN takes the role of a radio access server and the CN behaves as a client requesting access services from UTRAN. The same is true in the Iub interface, where BS is the server and its CRNC is a client. To some extent this applies also to the Iur interface, where actually the DRNC as a server provides the SRNC with a control service over the remote BSs. In the client–server based protocol design the behaviour of a server protocol entity is specified in terms of which actions it should take when receiving a service request from its client. On the other hand, under which conditions the client generates such service requests, may be left more unspecified.

The UTRAN protocol architecture is discussed in more details by Holma and Toskala (2001).

9.1.3. The CN Protocol Reference Model

The CN consists of the network elements, which provide support for the network features and end user services. The support provided includes functionality such as the management of user location information, control of network features and services and the transfer (switching and transmission) mechanisms for signalling and for user-generated information.

Within the CN the 3GPP R99 protocol architecture is derived from that of the GSM/GPRS system. Therefore five main protocol suites can be distinguished:

- Non-access stratum protocols between UEs and CN
- Network control signalling protocols between serving and home networks
- Packet data backbone network protocols
- Transit network control protocols
- Service control protocols

Due to the GSM/GPRS legacy these protocols have quite different specification backgrounds and origins.

First of all, the UMTS CN terminates the non-access stratum protocols from UEs. Among these the GSM/GPRS background is very evident. This is reflected by the 3GPP specification TS 24.008, which is an evolved version of its famous "ancestor" GSM 04.08. The design goal for the non-access stratum protocols has been to maintain compatibility between GSM/GPRS and UMTS systems in order to support combined networks (as shown in Figure 2.9) and dual-mode UEs.

The UMTS non-access stratum protocol stacks are shown in Figure 9.3. All of these protocols are carried over the signalling connection, which is established between UE and CN at the initial access and signalling connection establishment phase. In both PS and CS domains two sublayers can be distinguished in the non-access stratum protocols. The lower sublayer has to do with Mobility Management (MM); in the CS domain this protocol is called the MM protocol and in the PS domain it is called the GMM protocol to reflect the fact that it deals with the GPRS MM. On top of this common sublayer the more service specific Communication Management (CM) protocols are operated. The CM protocols and their control functions are as follows:

- Session Management (SM) protocol, which controls the establishment and release of packet data transfer sessions or Packet Data Protocol (PDP) contexts in the CN PS domain

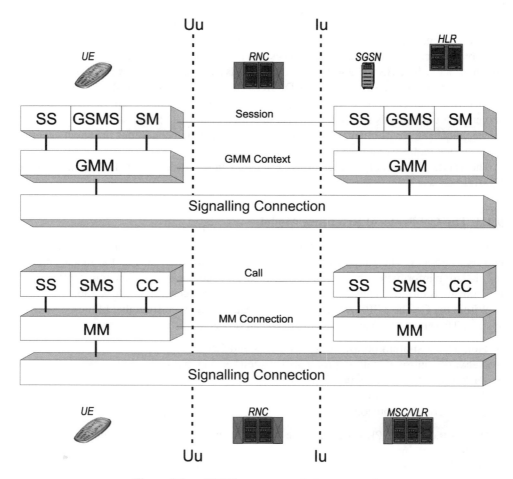

Figure 9.3.   UMTS non-access stratum protocols

- Call Control (CC) protocol, which controls the establishment and release of circuit switched calls in the CN CS domain
- Supplementary Service (SS) protocol, which controls the activation and deactivation of various call-related and non-call related supplementary services
- Short message service (GSMS/SMS) protocol, which controls the delivery of short text messages to and from UEs.

The network control signalling between serving and home networks utilise the Mobile Application Part (MAP) protocol suite, which was originally designed for controlling the GSM circuit switched services. With the introduction of packet switched services by the GPRS subsystem it became necessary to extend the MAP protocol to the interfaces between GPRS support nodes and the home network nodes. From the 3G CN interfaces, which were illustrated in Figure 5.4, the following ones are controlled by the 3G version of MAP protocol:

- Interfaces C, D, E, F and G, which originate from the GSM system
- Interfaces Gc, Gr, Gf and Gs, which originate from the GPRS system

This means that normally both SGSN and GGSN must be equipped with the MAP protocol implementation for control plane purposes.

The MAP protocol follows a transaction-oriented communication scheme. Each transaction (e.g. registering the subscriber's location at HLR) is executed as a dialogue between CN nodes. This communication structure is created for the MAP sublayer by the Transaction Capabilities Application Part (TCAP) sublayer below. The TCAP further on utilises an SS7 based signalling transport network as the backbone for signalling network across network operator and national boundaries.

The protocol suite for the packet data transfer within the CN PS domain has also been adopted from the GPRS backbone network. It follows the internetwork protocol paradigm and IP protocol suite, which had fully established itself by the time GPRS specification was started. Actually only one GPRS-specific protocol was needed to be added. This GPRS Tunnelling Protocol (GTP) controls the communication across the PS domain backbone network in interfaces Gn and Gp (Figure 5.4).

The GTP protocol can also be modelled as two subprotocols: GTP-C for control signalling, which operates between GGSN and SGSNs and the user plane part GTP-U which extends from GGSN across the Iu interface to the UTRAN side. This termination of GTP-U communication on RNC (instead of SGSN) differs from the original GPRS specifications.

Any UMTS network has to interwork with external telecommunications and data communications networks. This interworking takes place across the transit network, which in the case of CS services is the international PSTN/ISDN backbone or in the case of PS services most often another IP backbone network. The protocols in the transit network boundary must therefore be aligned to those used in the transit network backbone. In the telephone network case the ISDN User Part (ISUP) protocol is the obvious choice and correspondingly IPv4 is still the dominating backbone protocol in IP data networks.

Within the UMTS CN value-added services should be available to subscribers not only within their home networks but also in visited serving networks. Therefore the service control protocol CAMEL Application Part (CAP) has been carried over from the GSM infrastructure as described in network evolution in Chapter 2. The CAP protocol is another transaction-based protocol and can therefore use the TCAP service and international SS7 signalling backbone just like the MAP protocol discussed above.

9.2. UMTS Protocol Interworking Architecture

The number and wide variety of UMTS system protocols described in Section 9.1 above may look confusing – and due to the legacy considerations which had to be taken into account in the protocol selection and design, it cannot be taken as a uniform and homogeneous protocol suite. In this section a combined protocol architecture model for the UMTS network is elaborated in order to guide the later examination of individual protocols in this chapter.

While traversing through the radio interface, UTRAN and CN protocols, we have already familiarised ourselves with two major protocol design aspects:

- Separation of (generic) transport aspects from (UMTS-specific) mobile networking aspects by layering.

- Separation of network control aspects from the user data aspects by planes.

Before starting to elaborate from this basis, let's take a closer look at two additional fundamental design considerations in building network-wide protocol architectures:

- Protocol interworking
- Protocol termination

Protocol interworking deals with protocols, which belong to the same layer but extend across multiple network elements and therefore comprise a set of protocols, which contribute to (distributed) execution of a common system-wide function. As examples of such functions one might consider radio access bearer set-up or location updating. The first requires interworking from UE up to the CN element and co-ordinated actions from the RRC protocol and the AP protocols within UTRAN. For location updating the "chain" of interworking protocols consists of the MM protocol (between UE and CN) and the MAP protocol between the CN nodes. Such a chain of distributed actions is called a (system) procedure among the protocol designers.

Protocol interworking is often designed as an extension to the above mentioned client–server model. Once an event is triggered at one edge of the network the protocol entity first discovering this external event becomes the client, which starts the procedure by requesting some action from its server. In a UMTS network the server seldom can satisfy such a request alone, but has to assume a role of another client and request other services from its server(s). This is how the procedure goes on, until those servers – often at the other edge of the network than where the triggering event took place – are reached, which cannot delegate the task any further but have to obey the commands themselves. In the case of radio access bearer set-up the procedure is initiated by a CN element and finally executed by BS and UE, which set up the radio link between them. In the location updating it is the UE, which initiates the procedure and finally the HLR has to record the new location into its database.

More complete examples of UMTS system level procedures are shown in Chapter 10, but this interworking of protocols within a layer is the key design aspect in setting up a layered view on the UMTS protocols in this Chapter, also.

Protocol termination can now be taken as the end point of a protocol chain, which interworks in the manner described above. Within a network element, where a protocol terminates, a system function or algorithm can be identified as the source/sink of the data or commands. For the MM protocols mentioned above in the location update example the termination points are therefore UE and HLR. Termination is sometimes discussed with respect to a single protocol, but then the termination points are simply the two end-points of the peer-to-peer protocol. Therefore, from the system-wide protocol architecture point of view, the termination of a set of interworking protocols, is much more significant.

Throughout the rest of this chapter the UMTS protocols will be described by dividing the UMTS protocol suite into three different layered subsystems. Each of them may be seen as a logical network of its own. This decomposition neither follows the functional network architecture described in Chapter 1 nor the strict structure of the 3GPP specifications, but rather takes a network-wide protocol-oriented view, where layering and end-to-end interworking are the main structuring principles.

First the UMTS protocol model is divided – in horizontal decomposition – into layers. The protocols within each of the layers operate across multiple interfaces and belong together in the sense of interworking described above.

Three layers can be distinguished as shown in Figure 9.4:

- Transport network layer
- Radio network layer
- System network layer

In this division the transport network (layer) is responsible for providing a general-purpose transport service for all UMTS network elements thus making them capable of communicating across all the interfaces described earlier. The UMTS system functionality is then distributed among the network elements by using the protocols within radio and system network layers, which are by definition UMTS system-specific. The radio network (layer) protocols ensure interworking between UE and CN on all radio access bearer related aspects. The system network (layer) protocols extend from UE until the transit network edge of the UMTS CN. They ensure interworking on UMTS communication service related aspects.

Within all three layers it is then possible to distinguish between control aspects and user data transfer aspects, which create the other structuring dimension to the UMTS protocol model shown in Figure 9.4. All protocols dealing with control aspects only belong to the control plane and those dealing with user data transfer belong to the user plane. The distinction between control and user planes is most visible within the radio and system networks, but within the transport network many protocols are common to the control and user plane. For optimisation reasons some of the transport network protocols are however selected specially to carry control or user data only.

The control plane protocols in different UMTS system interfaces interwork with each other to ensure system-wide control of communication resources and services. In a

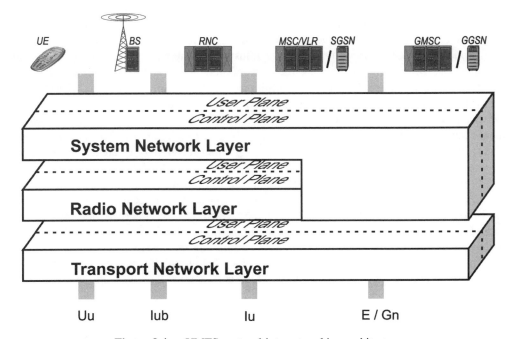

Figure 9.4. UMTS protocol internetworking architecture

similar way the user plane protocols interwork with each other to ensure the end-to-end flow of user data.

Based on the characteristics of the user traffic (within the CN) two different system domains are distinguished. In the Circuit Switched (CS) domain the user plane capacity is allocated in circuits for the whole duration of the service (e.g. a phone call). In the Packet Switched (PS) domain the user plane capacity is allocated in data packets. Such different requirements of these two domains are also reflected in the selection of transport network protocols and in the design of the system network protocols, which are therefore in this chapter often illustrated separately for both domains.

With this separation of functionalities the three different protocol subsystems can more easily be upgraded when new technologies become available. In fact such evolution is already taking shape in the introduction of alternative transport network protocols and the new IP Multimedia Subsystem (IMS) in release 4/5 of the 3GPP specifications.

9.3. Transport Network Protocol Aspects

Although ideally the communications capacity for user plane and control plane could be physically separate, it would lead to wasteful use of radio and terrestrial capacity in the network. Therefore the protocols providing general-purpose transport service for both user and control plane were designed as much as possible as a subsystem of its own, referred to as the transport network.

As the name suggests the UMTS transport network is actually a network within the UMTS network. A more detailed look at the transport network technology will further reveal that it is not just a single network, but actually consists of a number of networks, which when used in a UMTS system-specific way can been as a single (logical) UMTS transport network.

Close to the physical transmission of bits there is a transmission network, which except in the UMTS radio interface is typically based on digital transmission network technology. The (optical) trunk-line capacity of such a transmission network is shared by statistical multiplexing, which is controlled by the transmission network nodes. This transmission capacity is then utilised by the "second-degree" network on top of the transmission network. In the 3GPP R99 the preferred technology for this switching network is cell switching. This means that the data cells traverse through multiple switches before reaching their destination. Only at the edges of this switching network can we then find the actual UMTS network elements, which use this cell-switching technology to create the transport service between them.

The principle of composing an end-to-end transport network from a number of interconnected subnetworks is well known from the Internet protocol architecture. However, even if the UMTS transport network (in its 3GPP R99 level) is seen as an "internetwork", it does not follow such a uniform protocol structure as the IP internetworks. Instead the transport service within the different parts of the UMTS system – radio interface, RAN and CN – is designed and optimised for each part separately. As discussed below, there is a drive towards such a harmonised transport network architecture, which is based on IP networking with real-time transport capability.

Since the transport network itself provides switching and routing capabilities, it may also require an embedded control plane for the creation of transport network circuits or for distributing the routing information. These kinds of embedded control protocols are selected according to the transport network technology itself and they are not discussed in any more detail here.

9.3.1. Transport Network Protocol Architecture

The transport network consists of the lowest layers of the UMTS protocol architecture thus providing facilities to transport and route both control and user traffic across all UMTS network interfaces. Figure 9.5 presents the most important transport network configurations in 3GPP R99, which are signalling transport network and packet transport network.

Control Plane Protocols in Transport Network:

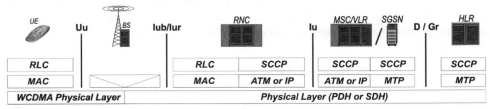

User Plane Protocols in Transport Network (PS Domain only):

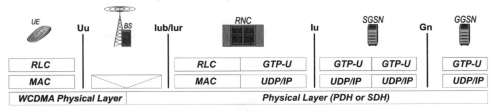

Figure 9.5. Transport network protocols in 3GPP R99

The physical layer controls the physical media through which both control signalling and user data traffic is transferred. The physical layer of the UMTS radio interface is based on the WCDMA radio technology as described in Chapter 3. The purpose of the physical layer protocol is to provide the upper layers with a set of WCDMA transport channels.

Over the terrestrial interfaces the physical layer may be formed by using general-purpose digital transmission technologies like Synchronous Digital Hierarchy (SDH) or Plesiochronous Digital Hierarchy (PDH). These are the most common alternatives available in transmission networks but in general any digital transmission system capable of providing stable bit flow with adequate bandwidth can be used. The SDH and PDH are the most obvious alternatives since they fulfil these basic requirements and are based on industrial standards.

The 3GPP release 1999 prefers ATM and its Adaptation Layers (AAL(n)) to be used especially in the UTRAN interfaces of the transport network protocol stack. The other recognised alternative, which is partly defined for release 1999 and is being standardised further in the release 4/5, is to use the IP suite over terrestrial interfaces.

The transport network protocol stack for the UTRAN radio interface as shown in Figure 9.5 is simplified a bit. Due to the UTRAN internal protocol structure and functional division the protocols MAC and RLC are terminated at the RNC. Therefore there is a need to carry the

MAC/RLC frames over the Iur/Iub interface. This part of the transport network is also based on ATM and its adaptation layers, but for simplicity reasons this is not shown in detail in Figure 9.5. Within UTRAN it is the SRNC, which is responsible for the radio interface related activities for a UE on the WCDMA transport channel level and the BSs actually only maintain the WCDMA physical channels.

9.3.2. Physical Layer Transmission Aspects

9.3.2.1. Physical Layer in the Uu Interface

The physical layer protocol controls the use of WCDMA physical channels (see Chapter 3) at the Uu interface (Figure 9.6). It provides the upper layer with a set of WCDMA transport channels, which characterise the different ways data can be transmitted over the WCDMA radio. The mapping between transport channels and physical channels is done by the physical layer protocol.

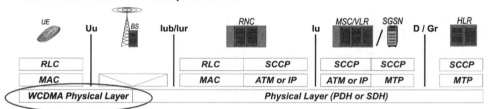

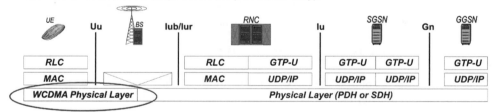

Figure 9.6.   Physical layer in Uu interface

The physical layer provides bandwidth-on-demand service, which means that the transport channels support variable bit rates. As the lowest layer, the physical layer is responsible for multiplexing at the WCDMA frame level. To manage this kind of multiplexing the physical layer controls the transport formats used on transport channels. The *transport format* is a format for the delivery of a transport block set during each transmission time interval on a transport channel. *Transport block sets* are the physical layer SDUs composed by the MAC protocol entities, which use the physical layer service.

The *transmission time interval* determines how often the physical layer can accept data from the MAC layer. The transport format attributes are listed in Table 9.1.

The physical layer implements the multiplexing of transport blocks from different transport channels. In order to maintain the frame synchronization of the WCDMA radio transmission, the physical layer also performs segmenting of transport blocks into radio frames. Other func-

Table 9.1   Transport format attributes for WCDMA-FDD

Dynamic attributes	Transport block size
	Transport block set size
Semi-static attributes	Transmission time interval
	Error protection scheme: type, coding rate, rate matching
	Size of CRC

tions of the physical layer are channel coding, interleaving and rate matching, which are all done before the radio frames are mapped for transmission over different physical channels.

As a comparison to the physical layer protocol in the GSM radio system it is worth noting that there is no ciphering at the WCDMA physical layer.

The physical layer protocol also attaches CRC checksum (0, 8, 12, 16, or 24 bits) into transport blocks to be transmitted over the radio interface. The receiving side the physical layer provides the transport blocks to the higher layer together with the possible error indication resulting from the CRC check.

Services provided by the WCDMA physical layer protocol are described in the 3GPP specification 25.302 and the detailed specification of the WCDMA physical layer protocol is covered by the 3GPP specifications TS 25.211–25.215. Detailed presentation of the WCDMA physical layer can be found in Holma and Toskala (2001).

9.3.2.2. The Physical Layer in Other Interfaces

In other, terrestrial, interfaces the physical layer can be implemented in many ways (Figure 9.7). The most obvious alternative is to utilise existing transmission, which typically consists

Control Plane Protocols in Transport Network:

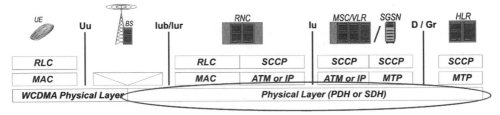

User Plane Protocols in Transport Network (PS Domain only):

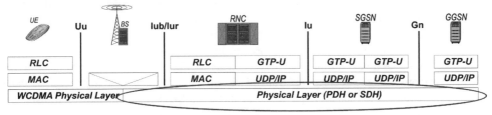

Figure 9.7.   Physical layer in other interfaces

of E1/T1 trunk lines with Pulse Code Modulation (PCM) technology. This is especially the case when UMTS is implemented on top of existing GSM CN. On the other hand, UMTS requires the transmission to be flexible and able to provide adequate and changeable bandwidth in a physical sense without wasting transmission resources.

E1 trunk lines use timeslots, which in turn form a fixed timing structure and at the same time it fixes the transmission capacities offered for various transactions. Thus, transmission arrangement based on pure E1 trunk lines is not the best possible solution. E1 trunks can be used but without the timeslot structure. If used like this, E1 trunks offer bandwidth and with suitable multiplexing arrangements the adequate bandwidth is formed over several E1 connections. This arrangement is called inverse multiplexing. For example, the Iub interface between one BS and RNC could consist of four E1 trunks using inverse multiplexing. In this case, the Iub bandwidth available could be 4×2.048 Mb/s $= 8.192$ Mb/s.

Actually, Figure 9.8 represents the lowest hierarchy level of PDH. In PDH the multiplexing always occurs in groups of four, where the next higher hierarchy level is always a combination of four previous levels. Thus, the (European) PDH hierarchy levels are 2, 8, 34 and 140 Mb/s. Transmission signals having greater than 140 Mb/s bit rates are not specified in PDH. In addition to this, fibre optics interfaces are out of scope of PDH.

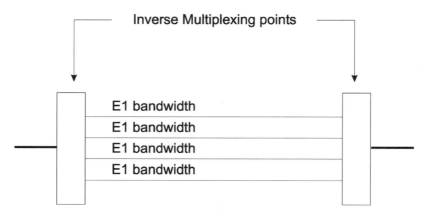

Figure 9.8.   Inverse multiplexing – principle

PDH is very much based on 2.048 Mb/s (2 Mb/s) signals; the bit rate is always increased in steps of 2.048 Mb/s and also the system is synchronised on a 2.048 Mb/s level. The equipment performing higher level multiplexing use their own, plesiochronous synchronisation signals. This actually gives the PDH system its name. The weak point of PDH is this synchronisation scenario. Because the end-to-end synchronisation occurs on a 2.048 Mb/s level, it is very difficult to separate single 2.048 Mb/s or 64 kb/s or any other bandwidths from the higher-level bit stream.

SDH (Synchronous Digital Hierarchy), as a technology fixes the limitations of the PDH technology. To make it very simple, the SDH is a way to offer end-to-end synchronisation for both synchronous and plesiochronous transmission simultaneously. For instance, the PDH signal can be transported as SDH payload. In this case SDH forms a kind of transparent "tube" carrying the PDH signal from one end to another and the PDH signal is available for demultiplexing in the destination.

The smallest transmission unit in SDH is STM-1 (Synchronous Transfer Module # 1). STM-1 is actually a frame structure defining the way to synchronously transfer bits with a bit rate of 155.52 Mb/s. The STM-1 frame consists of header(s) and payload. The header part of the STM-1 frame holds synchronisation information and error correction information. The payload part of the STM-1 frame carries information to be transferred. The next hierarchy level in SDH is STM-4 containing $4 \times$ STM-1 which is equal to a bit rate of 622.08 Mb/s. The third SDH frame structure is STM-16 which is equivalent to a bit rate of 2488.32 Mb/s.

From the network point of view, the SDH is a more flexible technology since it offers the possibility to extract a single signal/bandwidth out from the main stream without difficult demultiplexing arrangements. This is true if the signals the SDH carries are synchronous ones. If the SDH carries plesiochronous signals they cannot be directly handled like the synchronous ones.

9.3.3. Traffic Flow Multiplexing Aspects

In order to use the transmission capacity provided by the physical layer in the most effective way multiplexing of traffic from different sources must be done within the transport network. Besides the WCDMA physical layer, which is itself capable of some multiplexing of traffic onto the WCDMA physical channels, multiplexing is a task given to the layers above the physical layer in the transport network.

In this section the protocol-oriented mechanisms for multiplexing are considered first at the UTRAN radio interface and then within the UTRAN and to some extent across the CN also.

9.3.3.1. Transport Network Protocols in the Uu Interface

The transport network protocols, which carry all control signalling and user data – for both packet and circuit switched traffic – over the Uu interface are

- MAC protocol, and
- RLC protocol

These two protocols work together as the radio link of the transport network.

Medium Access Control (MAC) Protocol

The MAC protocol is active at UE, BS and RNC entities (Figure 9.9). MAC is a layer 2 protocol and as part of the common transport network it provides services to both control and user plane. In the design of the MAC protocol both circuit switched and packet switched traffic, as well as signalling traffic considerations, have been taken into account from the very beginning. This is a major difference from the 2G packet radio protocols, which can be seen as packet-data specific add-ons on top of the normal circuit switched (time-slotted) 2G radio carriers.

MAC provides its service as a set of logical channels, which are characterized by what type of data is transported on them. There are four kinds of control channels and two kinds of traffic channels among those logical channels as described in Chapter 3.

MAC has the overall responsibility of controlling the communications over the WCDMA transport channels provided by the physical layer. In order to be able to share the capacity of the transport channels among the set of users, the MAC protocol uses transport blocks as units

Control Plane Protocols in Transport Network:

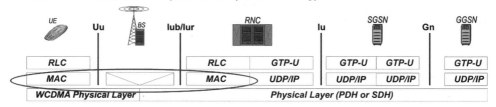

User Plane Protocols in Transport Network (PS Domain only):

Figure 9.9. Transport network – medium access control

of transmission. The possible transport block sizes for each transport channel are offered to MAC by the WCDMA physical layer as a part of transport format set definition

The MAC protocol is capable of multiplexing PDUs from higher layers into transport block sets carried over common transport channels. Identification of UEs on common transport channels is performed by a temporary identity (allocated by RNC), which can be either Cell-Radio Network Temporary Identity (C-RNTI, 16 bits) or UTRAN Radio Network Temporary Identity (U-RNTI, 32 bits).

On dedicated transport channels MAC is capable of multiplexing by composing transport block sets from the higher layer PDUs. In this case all blocks in the set must belong to a single UE for which the transport channel has been dedicated and multiplexing is possible only if the QoS parameters are identical for the services supported by the dedicated channel.

This combination of different transport formats is based on the instantaneous bit rate of the traffic offered at any moment together with the power control considerations and results in transport format combinations as illustrated in Figure 9.10.

Due to the fact that MAC is also part of the user plane, it is a real time protocol. MAC has to meet the tight timing requirements of the physical layer. MAC must be ready to send a new set of transport blocks according to the transmission time intervals indicated by the physical layer. During each time interval MAC must select the optimal transport format combination based on characteristics of the data to be transmitted and on current traffic situation.

The basic service of the MAC layer is to provide data transfer of MAC SDUs between peer MAC entities. From the upper layers point of view it is worth noticing that this service is given without acknowledgements and segmentation.

Switching between common, dedicated and shared transport channels is also executed by the MAC layer, although the commands to switch the configuration are coming from the RRC layer.

MAC also collects statistical information about the traffic to be used by the RRC layer. These MAC measurements include local measurements such as buffer occupancy, variance

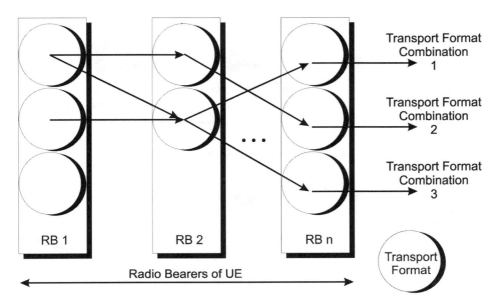

Figure 9.10.   Transport format combinations in MAC protocol

and average. Also MAC status indication about underflow or overflow detected for each transport channel and measurement mode (event triggered, periodic) are reported to the RRC.

The MAC layer performs a ciphering algorithm of transparent mode RLC data. The MAC SDUs send on dedicated logical channels; DCCH and DTCH are ciphered with a UE specific key by using the block cipher algorithm KASUMI as described in Chapter 8.

The MAC protocol has only one PDU called data-PDU, which consists of a MAC SDU and MAC protocol header. The header is constructed according to the transport channel to be used. In some cases, e.g. when the UE has only one dedicated channel, the MAC header may be left out.

The new work items proposed within 3GPP for MAC extensions in release 4/5 include support for High Speed Packet Access (HSPA).

The complete specification of the MAC protocol is given in 3GPP specification TS 25.321 and a more detailed description of the MAC protocol can be found in Holma and Toskala (2001).

Radio Link Control (RLC) Protocol

The RLC protocol runs both in RNC and UE and it implements the regular (data) link layer functionality over the WCDMA radio interface (Figure 9.11). The RLC is active both in control and user plane simultaneously and it provides data link services for both circuit switched and packet switched connections.

The operational environment of the RLC depends on the plane discussed; in the case of the user plane the RLC is used by the PDCP protocol and in the case of the control plane the RLC is used by the RRC protocol. The lower layer protocol for the RLC is MAC, which provides data transfer services as MAC SDUs over logical radio channels. Because the MAC data transfer mode is unacknowledged, the RLC becomes responsible for the delivery of such higher layer

Control Plane Protocols in Transport Network:

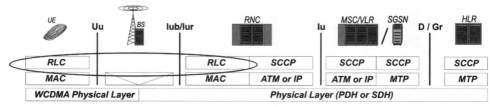

User Plane Protocols in Transport Network (PS Domain only):

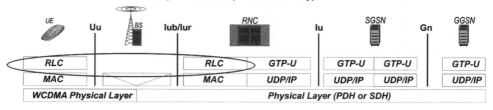

Figure 9.11.   Transport network – radio link control

PDUs, for which reliability is needed. Also, because the MAC layer is not able to segment larger SDUs according to the transport formats available, RLC takes care of such functionality. Therefore the transport formats are visible at the RLC layer as shown in Figure 9.12.

RLC provides the data transfer service of higher layer PDUs as RLC SDUs. This service is called radio bearer service. Three modes of operation have been defined for RLC: transparent (Tr), unacknowledged (UM) and acknowledged (AM).

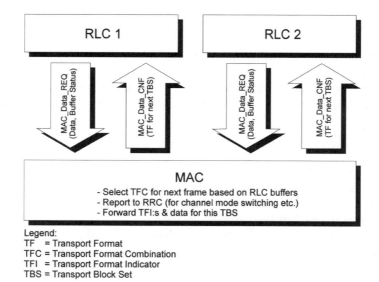

Figure 9.12.   RLC interoperation with MAC on dedicated channels

In the *transparent mode* the RLC transmits SDUs without adding any protocol information. Segmentation and reassembly is still possible, but it has to be negotiated at the RRC layer during radio bearer set-up. This mode is used, e.g. for streaming class services.

In the *unacknowledged mode* the RLC transmits SDUs without guaranteeing delivery to the peer entity. This mode is used by some RRC control procedures, where acknowledgement is taken care of by the RRC protocol itself.

In the *acknowledged mode* the RLC transmits SDUs and guarantees the delivery to its peer entity. The guaranteed delivery is ensured by means of retransmission. In case RLC is unable to deliver the data correctly, the user of RLC at the transmitting side is notified. This mode is used for packet switched data transfer over dedicated logical channels.

The RLC has numerous functions to perform on its layer such as:

- Segmentation and reassembly
- Concatenation
- Padding
- Error correction
- In-sequence delivery of SDUs
- Duplicate detection
- Flow control
- Sequence number check
- Protocol error detection and recovery
- Suspend/resume functionality
- SDU discard
- Ciphering

Segmentation and reassembly means the adjusting of variable-length higher layer PDUs into/from RLC PDUs. The RLC PDU size is adjustable to the actual set of transport formats.

Concatenation is used when the contents of an RLC SDU do not fill an integer number of RLC PDUs. Then the first segment of the next RLC SDU may be put into the RLC PDU in concatenation with the last segment of the previous RLC SDU. When concatenation is not applied and the remaining data to be transmitted does not fill an entire RLC PDU of a given size, the remainder of the data field shall be filled with padding bits. Padding can be replaced by piggybacked status information for the reverse link.

Many functions of RLC have to do with error detection and correction. The bit errors may actually be detected by the physical layer CRC check, but the RLC layer is responsible of error recovery. The most effective error recovery is provided by retransmission in acknowledged data transfer mode. The RLC protocol can be configured by RRC to obey different retransmission schemes, e.g. Selective Repeat, Go Back N or Stop-and-Wait.

In-sequence delivery of SDUs preserves the order of higher layer PDUs that were submitted by using the acknowledged data transfer service. If this function is not used, out-of-sequence delivery may happen.

The RLC is able to detect duplicates of the received RLC PDUs and can ensure that the resultant higher layer PDU is delivered only once to the upper layer. Flow control allows an RLC receiver to control the rate at which the peer RLC entity may send information.

Sequence number check function may be used in unacknowledged mode to guarantee the integrity of reassembled PDUs. It provides a mechanism for the detection of corrupted RLC

SDUs through checking the sequence number in RLC PDUs when they are reassembled into a RLC SDU. A corrupted RLC SDU will be discarded.

Protocol error detection and recovery functionality detects and recovers from errors in the operation of the RLC protocol. The RLC reset procedure is used to recover from an error situation. In case of an unrecoverable error the RLC entity notifies the RRC layer.

The RLC may suspend and resume the data transfer under request from the RRC.

The SDU discard function allows the discarding of the remaining RLC PDUs of that SDU from the buffer on the transmitter side, when the transmission of an RLC PDU does not succeed for a long time. The SDU discard function allows the avoiding of buffer overflow. There are several alternative operation modes of the RLC SDU discard function.

Ciphering is performed in the RLC layer for those radio bearers, which use unacknowledged or acknowledged mode. Ciphering is done with a UE specific key by using the block cipher algorithm KASUMI as described in Chapter 8.

The complete specification of the RLC protocol is given in 3GPP specification TS 25.322 and a more detailed description of the RLC protocol can be found in Holma and Toskala (2001).

9.3.3.2. Transport Network Protocols in Other Interfaces

Unlike in the radio interface the transport network protocols for the UMTS terrestrial interfaces have not been designed specifically for UMTS transport purposes, but instead the 3GPP standardisation bodies have selected them among the existing protocol suites. As in the case of all transport network protocols the focus in this selection has been on the ability to multiplex traffic from different end users and – to be more precise – from different UMTS bearers taking into account their QoS characteristics.

This – together with some standardisation politics, which inevitably is also present – has led to the selection of two major protocol suites: one coming from the broadband telecommunications, the ATM protocol family, and the other coming from the Internet data communication networks and known as the IP protocol family. In the 3GPP release 1999 the ATM is the dominating transport network technology on the UTRAN side and the same is true for IP-protocol technology in the PS domain on the CN side. Within the CN CS domain the transport network will continue to be based on the time-slotted PCM trunking network.

In order to make the ATM and IP protocols suitable for the UMTS transport network service, both of them have required some adaptations as compared to their use in their native networking environments. In the case of ATM such adaptations mostly have to do with conversational speech transmission capability whereas the IP protocol stack has been extended to meet the multiplexing of user data and signalling transport requirements. The single most significant extension to the IP transport is adoption of the GPRS tunnelling protocol to carry user data packets between UTRAN and CN nodes.

ATM Transport

The basic idea in ATM (Figure 9.13) is to split the information flow to be transferred into small pieces (packets), attach address tags to those packets and then transfer the packets through the physical transmission path. The receiving end collects received packets and forms original-like information flow from the contents of the packets. The packet containing transmitted information is called the *ATM cell*.

Control Plane Protocols in Transport Network:

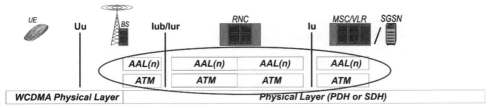

User Plane Protocols in Transport Network (PS Domain only):

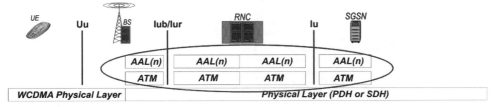

Figure 9.13. Transport network – asynchronous transfer mode

An ATM cell as shown in Figure 9.14 consists of two parts, a 5-byte-long header (address information) and payload (transmitted information). When comparing to "conventional" protocols and messages, the header is very short. This sets some limitations on what can be done but on the other hand, the information transfer effectiveness is high: the addressing overhead is $5/(5 + 48) \approx 9.5\%$. Another aim has been to establish a very lightweight transmission system without any extra "bureaucracy". Because of this, the payload of an ATM cell is *not* protected with check sum method(s). Nowadays this is possible because the transmission networks carrying ATM traffic are of high quality and the terminals used are able to perform error protection themselves if required.

Figure 9.14. ATM cell structure

The header of an ATM cell contains some address information as illustrated in Figure 9.15. The most essential items are:

- VPI (Virtual Path Identifier): the identifier for a Virtual Path (VP), or more generally, an identifier for a constantly allocated semi-permanent connection.
- VCI (Virtual Channel Identifier): an identifier for a Virtual Circuit (VC). This field is long because there may be thousands of channels to be identified within one VP. For instance, multimedia applications may require several VCIs simultaneously, one VC per each multimedia component.

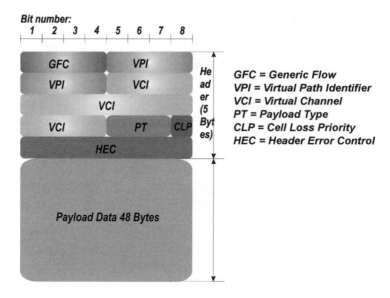

Figure 9.15.   ATM cell header structure

- PT (Payload Type): this indicates whether the 48-byte payload field carries user data or control data.
- CLP (Cell Loss Priority): this is a flag indicating if this ATM cell is "important" or "less important". If CLP = 1 (low priority/less important) the system may lose this ATM cell if it has to.
- HEC (Header Error Control): in ATM, the ATM cell header is error protected.
- Reason: a failure in the ATM cell header is more serious than in payload because due to header error the ATM cell may be delivered to the wrong address, for instance. The error correction mechanism used is able to detect all errors in the header and one failure can be corrected.

Figure 9.16 illustrates the transmission path of ATM. One ATM transmission path may consist of several virtual paths, which further on contain virtual channels.

A virtual path is a semi-permanent connection simultaneously handling many virtual connections/channels. Actual data is transferred in ATM cells over the virtual channels.

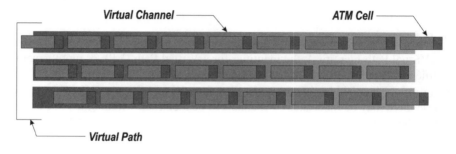

Figure 9.16.   Virtual Path (VP) and Virtual Channel (VC)

From the point of view of the UMTS system, an ATM transmission path is, say, between the BSs and the RNC. If a loop transmission is in question, the transmission path contains many virtual paths (one per BS) and the virtual channels in the virtual path are set up on a per call basis. The bandwidth of the virtual channel varies depending on the bearer service used.

The ATM layer as such consists of fairly simple transport media and in theory suitable for transmission purposes as such. In practise, the ATM layer must be adapted to the higher protocol layers and the lower physical layer. ITU-T has defined what are called ATM service classes with AALs.. The original idea was that each service class from A to D should correspond to one AAL from 1 to 4. As time went by, the original idea disappeared as can be seen from Figure 9.17.

The service classes of the ATM are

- Constant Bit Rate Service (CBR)
- Unspecified Bit Rate Service (UBR)
- Available Bit Rate Service (ABR)
- Variable Bit Rate Service (VBR)

Any transparent data transfer may use the CBR and the resources are allocated on a peak data rate basis. The UBR uses free bandwidth when available. If there are no resources available, queuing may occur. The ABR is used when the user service has a minimum bit rate defined. Otherwise the bandwidth is used as in UBR. The VBR provides a variable bit rate based on statistical traffic management.

The adaptation layers for ATM are as follows:

- AAL1 offers synchronous mode, connection oriented connection and constant bit rate for the services requiring this kind of adaptation.

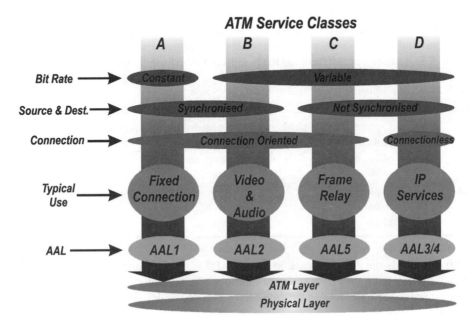

Figure 9.17. ATM Adaptation Layers (AAL)

- AAL2 offers synchronous mode, connection oriented connection with variable bit rate for the service using this adaptation.
- AAL3/4 offers connectionless and asynchronous connection with variable bit rate.
- AAL5 offers asynchronous mode, connection oriented connection with variable bit rate.

As shown in Figure 9.18, AAL is divided into two sublayers, Convergence Sublayer (CS) and Segmentation And Re-assembly (SAR) sublayer. The CS sublayer adapts AAL to the upper protocol layers and the SAR splits data to be transmitted into suitable payload pieces and in receiving direction it collects payload pieces and assembles them back to original dataflow. Depending on the case, the CS sublayer may be divided further into smaller entities.

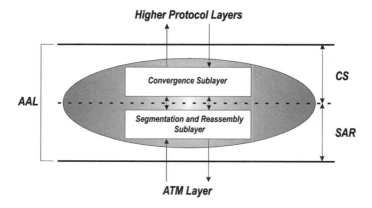

Figure 9.18.   General structure of AAL

From the point of view of the UMTS transport network, AAL2 and AAL5 are the most interesting alternatives. AAL2 is seen as a suitable option for circuit switched user plane connections and AAL5 is seen as suitable for control protocol exchange. A more detailed list of these AAL variants in different UTRAN protocol interfaces is given in Table 9.2.

Table 9.2   ATM adaptation layers used at the UTRAN interfaces

AAL5	Iu: CS C-plane, PS C/U-plane, internal C-plane of the transport network
	Iur: C-plane, internal C-plane of the transport network
	Iub: C-plane, internal C-plane of the transport network
AAL2	Iu: CS U-plane
	Iur: U-plane
	Iub: U-plane

Within UTRAN the ATM adaptation layer AAL5 is used to carry all control protocols as well as the PS domain user data at the Iu interface. On the other hand AAL2 is used as the common carrier for user data in all interfaces, except the PS domain user data at the Iu interface.

The sublayers of the AAL protocol stack in different UTRAN interfaces are not described in detail here, but can be found in the 3GPP specifications numbered in Figure 9.2. The key observation in the selection of these convergence protocols is that in all control plane protocols for the PS domain the convergence may be accomplished in two alternative ways. Either the convergence protocols specified by ITU-T for signalling transport over AAL5 are selected or IP-over-AAL5 is used. Since also the U-plane convergence for the PS domain at the Iu interface is based on User Datagram Protocol (UDP) over IP, this means that for the PS domain the transport network can be completely based on an "IP-over-ATM" implementation option. This serves as a migration path towards wider utilisation of IP-based transport in the 3GPP release 4/5.

IP Transport for User Data (GTP-U)

Within the 3GPP R99 set of specifications IP-based transport is widely applied only within the CN PS domain backbone network and at the Iu interface for PS domain user plane traffic as shown in Figure 9.19. The major protocol providing the transport network service in these interfaces is the GPRS Tunnelling Protocol for packet data User Plane (GTP-U). As the name suggests this protocol has been adopted from the 2G GSM/GPRS system, but with a significantly different architectural design choice. In 2G the GTP-U protocol is used only between the 2G-GSN network elements, but in 3G is was extended to reach RNC across the Iu–PS interface also.

Control Plane Protocols in Transport Network:

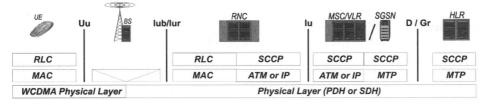

User Plane Protocols in Transport Network (PS Domain only):

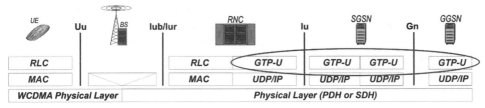

Figure 9.19.   GPRS tunnelling protocol for user plane

GTP-U is a transport network protocol supporting user plane data transfer in the UMTS network. GTP-U belongs to the packet switched domain in the UMTS network and is located in RNC on the UTRAN side, as well as in SGSN and GGSN on the CN side.

GTP-U operates on three different interfaces; Gn, Gp and Iu-PS. The Gn interface is between GSNs (i.e. SGSNs and GGSNs) belonging to the same UMTS packet domain network. The Gp interface is used in network interworking between two SGSNs or between SGSN and GGSN belonging to different UMTS packet domain networks. Iu-PS is the interface between RNC and SGSN.

On all interfaces GTP-U operates on top of the UDP/IP protocol family. User Datagram Protocol (UDP) provides connectionless message transfer between two IP network nodes. UDP addressing is used to identify GTP-U end points within source and destination network elements. UDP also provides checksum mechanism to detect transmission errors inside data packets. As it is generally known the main service of the IP is the routing of messages between source and destination network elements. This routing is based on IP network addresses (both IPv4 and IPv6 may be used within the UMTS transport network).

GTP-U provides connectionless data transfer services for upper layers. GTP-U allows multi-protocol user data packets to be tunnelled across the Iu-PS, Gn and Gp interfaces. Because of the encapsulation and tunnelling mechanisms it is possible to transfer user data packets utilising different routing protocols in the packet domain IP backbone network. Also UMTS network elements and protocols transferring user data packets between UTRAN and GGSN need not be aware of different addressing mechanisms at the IP layer in order to be able to associate user data packets to specific PDP contexts.

Since the main purpose of the GTP-U protocol is to transfer user data, it has been optimised to that specific task. GTP-U multiplexes packets received from one interface (e.g. Gi at the GGSN) and addressed to several destinations (different UEs). GTP-U receives the user data packet from the external packet network, interprets the destination of the packet and passes the packet onto the next node along the path. Tasks for setting up the tunnels between GTP-U endpoints have been excluded from GTP-U and control plane protocols are used for those purposes. On the Gn interface GTP-C (GTP for control plane) and on the Iu-PS interface RANAP protocols are used to control the set-up of GTP-U tunnels.

GTP-U protocol entities have to interwork with other protocol entities along the way the data packets are transported. In GGSN the GTP-U protocol entity communicates with the packet relay functionality located at the edge of the UMTS network, i.e. the Gi interface. At the Iu-PS side of an SGSN the GTP-U protocol entity interworks with another GTP-U entity at the Gn side. Within RNC the GTP-U protocol entity terminates the tunnel and forwards the packets to the frame protocol entity at the radio network layer.

The interworking between GTP-U and the neighbouring protocol entities is quite simple. After all, all parameters related to user data packet transfer have been negotiated in advance by the control plane. User data packets are just passed to the next entity, which is then able to handle the packet according to its predefined PDP context parameters.

The main functions of the GTP-U protocol are

- Data packet transfer
- Encapsulation and tunnelling
- Data packet sequencing
- Path alive check

The maximum size of the user data packet is 1500 octets. User data packets containing 1500 octets or less shall be transmitted as one packet. In case the user data packet received from the external network by GGSN exceeds the 1500 octet limit, GGSN shall fragment or discard the user data packet depending on the PDP type. Since IP fragmentation is inefficient and does not tolerate transmission errors, fragmentation should be avoided. Thus all links between network elements containing GTP-U should have Maximum Transmission Unit (MTU) values exceeding the 1500 octets plus size of GTP-U, UDP and IP headers.

Each user data packet is encapsulated before being transmitted into the UMTS packet domain network. The encapsulation adds a GTP protocol header containing tunnelling information into every user data packet. Tunnelling information includes a 32-bit Tunnel Endpoint Identifier (TEID), which serves two important purposes. Firstly, TEID is used to address a PDP context inside a tunnel endpoint. Hence, it also indirectly refers to different UEs, which have one or more PDP contexts active in the UMTS network. Secondly, tunnelling enables multiplexing of user data packets destined for different addresses into a single path that is identified by two IP addresses. This removes the need to understand different routing protocols the user may use in the UMTS network and thus makes the UMTS network simpler. GTP-U tunnelling makes it also simpler to allow end users to utilise new routing protocols.

As an optional feature GTP-U may preserve the order of user data packets between RNC and GGSN. During PDP context establishment at the control plane the reordering may be negotiated. In case user data packet reordering is applied, the GTP header shall contain a 16-bit sequence number, which GTP-U uses to determine if user data packets are received in the correct order or whether it has to wait for the missing packets. However, in case the missing user data packet is lost, the data packet sequencing mechanism does not provide a mechanism to recover a lost packet. It is up to the other layers to recover from the lost user data packet.

GTP may check at regular time intervals if the peer GTP is alive by sending an echo request message to the peer GTP. The peer GTP shall respond to an echo request message with an echo response message. On the Gn interface one may think that the path alive check procedure belongs to GTP-C but on the Iu-PS interface GTP-U is the only GTP element and thus handles the path alive check procedure. On the Gp interface the path alive check is similar to the Gn interface.

The main procedures of the UMTS network where the GTP-U protocol is used are first of all the user data packet transfer between UTRAN and SGSN as well as between SGSN and GGSN. When a routing area update procedure requires to switch a GTP-U tunnel from one SGSN to another, the GTP-U is used to relay user data packets not yet sent to the UE from the old SGSN to the new SGSN. GTP-U is also used as part of the SRNS relocation procedure to tunnel user data packets not yet sent to the UE from the source RNC to the target RNC via the pair of SGSNs attached to the RNCs.

The GTP-U protocol is defined in the 3GPP specification TS 29.060.

9.3.4. Signalling Transport Within the CN

Due to the inherited background, i.e. GSM and its elements and functionalities, the 3GPP R99 implementation still includes a protocol stack based on the Common Channel Signalling System 7 (SS7). SS7 is used for signalling transport within the CS domain of the CN as illustrated in Figure 9.20.

The SS7 protocol stack is divided into a physical layer, signalling link layer (MTP2) and signalling network layer (MTP3 and SCCP). The SS7 physical layer is based on PCM transmission, which makes it possible to multiplex user data (circuit switched voice) and signalling in a time-slotted manner. One of the 64 kb/s time-slots on a PCM trunk line is typically reserved for the signalling traffic. Besides this structure originating from the narrowband SS7 a broadband version of the protocol stack is also used within UTRAN. In this case MTP3 is run on top of the ATM protocol stack, which leads to a new adaptation of this protocol called MTP3b.

Control Plane Protocols in Transport Network:

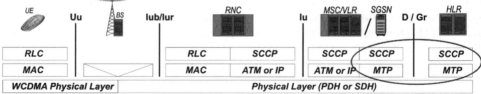

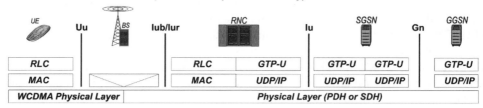

User Plane Protocols in Transport Network (PS Domain only):

Figure 9.20.   SS7 Transport in CN

The SS7 basic protocol stack provides signalling connections, their control and basic signalling routing functionality. For various application-oriented signalling needs one must add "intelligence" by adding more protocols to the protocol stack. For example, to offer connection-oriented and connectionless services within an SS7 environment, an additional protocol called Signalling Connection Control Part (SCCP) is required and this protocol sits on top of the basic SS7 protocol stack.

The SS7 uses three kinds of messages: FISU (Fill-In Signalling Unit), LSSU (Line Status Signalling Unit) and MSU (Message Signalling Unit). The signalling channel must be populated all the time, hence, if there isn't any information to send, the signalling node sends FISU. FISU does not contain any upper layer information, it only contains sequence number and indicator bits for acknowledgement purposes. LSSU is sent when the SS7 nodes need to negotiate/change a signalling channel status or they have to inform each other about other maintenance activities. MSU is sent when there is some upper layer information to be delivered.

As illustrated in Figure 9.21 the SS7 message is always started with a Frame Mark (*F*). *F* is a fixed bit pattern 01111110 and after this the receiving signal node is expecting some SS7 information to be received. After *F* the SS7 message contains sequence numbers in both back and forward directions (BSN and FSN) and indicator bits for the same directions (BIBand FIB). With these four fields the receiving node is able to perform message acknowledgement activities. In SS7 any message can acknowledge another; for instance a FISU can acknowledge an MSU.

The Length Indicator (LI) field indicates how many octets long the SS7 message is. Up to this point, all the fields in the message are related to data link layer activities.

The next part, Service Information Octet (SIO), indicates the user (protocol) to which this message is addressed. SIO may contain a bit patter indicating that, for instance, ISUP is the protocol handling this MSU. The real, higher-level message is in the Service Information Field (SIF). At the beginning of this field the MSU contains addressing information for the

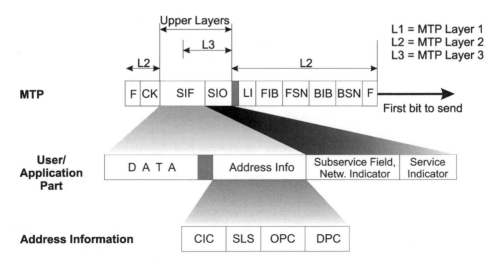

Figure 9.21. SS7 Message Structure (MSU)

MTP layer, which is responsible for message routing. In the SS7 network addresses of the signalling points are given as Signalling Point Codes (SPC). The MTP routing facility checks the Originating Signalling Point Code (OPC) and Destination Signalling Point Code (DPC) and if the DPC is the same as that defined for the current MTP node, the message is interpreted to be terminated in this signalling point. If different, the message is re-routed towards the correct DPC.

In addition, the SIF address part contains identification for used signalling channels, Signalling Link Selection (SLS) and circuit concerned, Circuit Identification Code (CIC).

The basic element of SS7 signalling information transfer is signalling link. Signalling link is a data link layer connection between two signalling nodes as shown in Figure 9.22. Both nodes identify a signalling link with a unique number, Signalling Link Code (SLC). The SLC (or SLS) should be same in both ends of the signalling link and checking for this is one of the items the LSSU messages are used for.

Figure 9.22. SS7 signalling link

One signalling link between two nodes is able to handle a certain amount of signalling traffic but sooner or later more links will be required. The set of signalling links between two signalling nodes, as shown in Figure 9.23, is called Signalling Link Set (SLS). For proper signalling link selection, every signalling link within a SLS must have unique SLC. The signalling traffic is carried through all the signalling links within the SLS.

Normal practise is that a signalling session (for instance signalling related to ISUP set-up) is carried through by using the same signalling link for all messages. If *load sharing* is

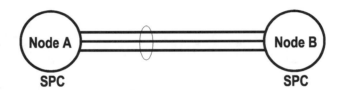

Figure 9.23.   SS7 signalling link set

used, the MTP level is able to distribute messages from one signalling session over several links.

The SS7 makes it possible to have a situation where the actual traffic path is geographically different from that of the related signalling. This is, two nodes may have direct traffic connections, but the signalling related to those connections is handled through other nodes.

The signalling node taking care of re-routing of the messages in this case is called STP (Signalling Transfer Point), please refer to Figure 9.24. In the originating signalling node, the routing entity on the MTP level is called Signalling Route Set (SRS). SRS is the collection of the signalling routes through which a certain SPC can be reached. The signalling route is in practise the same as the SLS, but the difference here is that SLS is not "aware" of the STP facility, but the signalling route is.

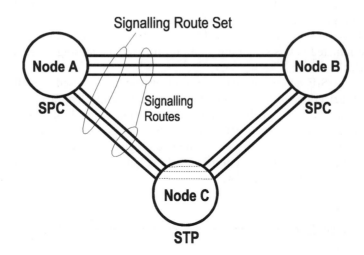

Figure 9.24.   SS7 signalling route set

The STP nodes are capable of routing signalling messages only within a single SPC addressing space, which is typically administered by a single operator. In order to make it possible to route signalling across network boundaries the SCCP protocol is needed. SCCP uses global title addressing, which makes it possible, for example, for an VLR within a visited network to reach the HLR in the subscriber's home network by using the GT of the HLR for addressing.

9.3.5. IP Option for Signalling Transport

As described in the content of GTP-U above the user data within the PS domain backbone network and across the Iu–PS interface is transported with the IP-based protocol stack. Besides these two obvious cases the IP-based transport started to gain more momentum towards the end of the 3GPP R99 specification campaign. In order to create a basis for full-scale utilization of IP internetworking technology as a common transport for the future evolution of UMTS networks, it became necessary to study the IP option for signalling transport as well (Figure 9.25).

Control Plane Protocols in Transport Network:

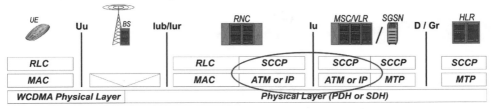

User Plane Protocols in Transport Network (PS Domain only):

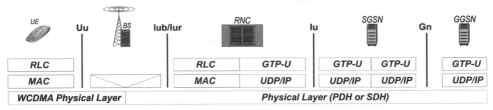

Figure 9.25.   Transport network – signalling transport in UTRAN

In order to maintain the harmonization between the IP transport and ATM transport options both stacks should provide the same transport service. Therefore it was decided to "surface" both stacks with the SCCP protocol entity, which then always delivers a harmonized signalling transport service. The detailed structure of the resulting protocol stack is illustrated in Figure 9.26.

The left hand side of the stack has already been discussed in association with the ATM protocols above.

The adaptation of the SCCP protocol on top of the IP protocol stack was achieved by using two convergence protocols, which are developed and standardized by the SIGTRAN Working Group in IETF. These protocols and the corresponding IETF documents are

- Stream Control Transport Protocol (SCTP), RFC 2960
- MTP3 User Adaptation Layer (M3UA) (Internet draft document)

The purpose of these two sublayers is to provide the SCCP layer with an illusion of sitting on top of the regular MTP3 service as it has been in the telecommunication signalling networks for more than two decades.

In the 3GPP R99 transport network the SCTP protocol entities can be found in RNC and

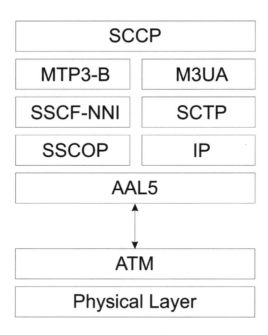

Figure 9.26.   Two options for signalling transport stack

SGSN only. This transport option is available in the Iu–PS interface for both control plane and user plane as was targeted from the IP-evolution point of view.

SCTP assumes that it is running over an IPv4 or IPv6 network. Even more important is that SCTP assumes it is running over a well-engineered IP network. This, in practice, means that there is a diverse routing network underneath so as to avoid a single point of failure. It takes into consideration multi-homed endpoints, which are endpoints with more than one IP address/port number tuple, for additional reliability. Further more, it provides a MTU discovery function, to determine the MTU size of the data path, to avoid IP level fragmentation.

The purpose of the SCTP protocol is to provide a robust and reliable transport signalling bearer. To achieve this SCTP provides appropriate congestion control procedures, fast retransmit in the case of message loss and enhanced reliability. It also provides additional security against blind attacks, which will be used to increase security in connecting together UMTS networks of different operators.

The operational environment of the M3UA protocol is the same as that of the SCTP protocol. It is running on top of the SCTP protocol, which provides a reliable transport bearer to M3UA.

The purpose of the M3UA protocol is to support SCCP signalling, so that the SCCP lower interface does not need to be modified. M3UA needs to manage the use of SCTP streams and to meet equal performance as its SS7 counterpart, the MTP3b protocol, while running on top of an IP stack. M3UA provides mapping of SS7 addressing to IP addressing. It provides failover support as well as the ability to loadshare among endpoints.

M3UA has two architectural modes: signalling gateway to IP Signalling Point (IPSP) and IPSP-to-IPSP. It is assumed that the IPSP-to-IPSP is the most likely model and the simplest as well.

9.4. Radio Network Protocols

In the UMTS internetwork protocol architecture (see Figure 9.4) the radio network protocols compose the next layer on top of the generic transport network protocols discussed above. The radio network extends from UE across the whole UTRAN and terminates at the Iu edge nodes of the CN. The protocols in the radio network are needed to control the establishment, maintenance and release of radio access bearers and to transfer user data between UE and CN along those radio access bearers.

9.4.1. Radio Network Control Plane

The control plane protocols in the radio network layer execute all control needed for management of radio access bearers. According to the client–server approach applied in the overall design of the UTRAN control plane these protocols implement the radio resource management decisions and maintain the RABs for all UEs even when they are moving within the UTRAN area.

9.4.1.1. RANAP – Radio Access Network Application Protocol

The Iu interface connects the UMTS radio access network and the CN together. The same Iu interface is used to connect both CS service and PS service domains to the UTRAN (see Figure 9.27). The Iu interface has been designed to support the independent evolution of the UTRAN and CN technologies. In addition to that, it was made possible to develop independently CS and PS service domains inside the CN and still use the same Iu interface to connect the domains to the UTRAN.

From a UTRAN perspective it is desirable that the connections to the CS and PS service domains are as similar as possible. That is why a single signalling protocol between the

Control Plane Protocols in Radio Network:

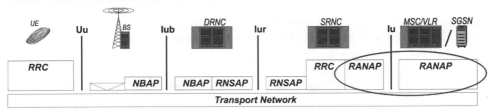

User Plane Protocols in Radio Network (PS Domain only):

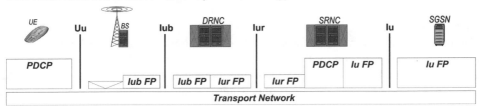

Figure 9.27.   Radio network – RANAP

UTRAN and CN was defined. This protocol is called Radio Access Network Application Part (RANAP) and it is defined in the 3GPP specification TS 25.413. Both CS and PS domains use the RANAP protocol to access the services provided by the UTRAN.

RANAP is the protocol that controls the resources in the Iu interface. One RANAP entity resides in the RNC and the other peer entity in the MSC or the SGSN.

RANAP is located on top of the Iu signalling transport layers. RANAP uses the signalling transport service to transfer RANAP messages over the Iu interface. In 3GPP R99 the transport layers in the Iu interface are comprised of an SS7 protocol stack over ATM or IP over ATM.

The SS7 stack consists of several protocol layers but the topmost protocol is always SCCP. In spite of the two possible transport options (SS7 or IP) it is always an SCCP protocol entity, which offers RANAP the interface to the services of the transport protocol stack as was described in Section 9.3.5

RANAP has certain requirements for the transport layers below. Basically it is assumed that each message (PDU), which RANAP sends to the peer entity will reach the destination without any errors. In other words it is the responsibility of transport protocol stack to provide reliable data transfer across the Iu interface for the RANAP PDUs.

Each PDU of the dedicated control services should be sent on a unique signalling connection. In the Iu interface, signalling connection is realised by an Iu signalling bearer. The transport layers shall provide RANAP the means to dynamically establish and release signalling bearers for the Iu interface when RANAP requests it. Each active UE shall have its own Iu signalling bearer. It is the responsibility of the transport layers to maintain the bearers. If the Iu signalling bearer connection breaks for some reason, SCCP informs RANAP about it.

RANAP offers services to the upper layer protocols: on the MSC/VLR and SGSN side the non-access stratum protocols make use of the RANAP services. By interworking with the RRC protocol in RNC, RANAP contributes to the transfer of signalling messages of these upper layer protocols between the UE and CN.

RANAP services are divided into two groups, general control services and dedicated control services. General control services are related to the whole Iu interface between the RNC and CN. Dedicated control services support the separation of each UE in the Iu interface: they are always related to a single UE. The majority of RANAP services are dedicated services. All the RANAP messages of the dedicated control services are sent on a dedicated connection. This connection is called a signalling connection.

The overall RAB management is one of the main services RANAP offers. RANAP provides the means for the CN to control the establishment, modification, and release of the RABs between the UE and CN. Related to that, RANAP supports UE mobility by transferring RAB to a new RNS when the UE moves from the area of the serving RNS to another. This service is called SRNS relocation. Controlling the security mode in the UTRAN is one of the RANAP services as well. All the above-mentioned services are dedicated control services.

The general control services are needed only in exceptional situations. RANAP provides, for example, a means to control the overload in the Iu interface if the amount of the user traffic grows too high. In the case of a fatal failure in either end of the Iu interface RANAP offers a reset service. That initialises the whole Iu interface and clears all the active connections.

The detailed specification for the RANAP protocol is in the 3GPP specification TS 25.413.

9.4.1.2. RNSAP – Radio Network Subsystem Application Protocol

RNSAP is a control plane protocol at the radio network layer (Figure 9.28). It provides control signalling exchange across the Iur interface. RNSAP is run by two RNCs, one of which takes the role of Serving RNC (SRNC) and the other acts as a Drifting RNC (DRNC). SRNC is the one having the RANAP signalling connection to the CN.

Control Plane Protocols in Radio Network:

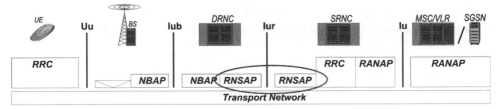

User Plane Protocols in Radio Network (PS Domain only):

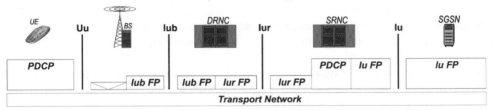

Figure 9.28.   Radio network – RNSAP

The RNSAP protocol operates always on top of the SCCP protocol. Like in the case of the RANAP protocol there are two signalling transport options below SCCP: either ATM with an AAL5 adaptation layer or IP-based transport, where SCTP and M3UA protocols are used as the adaptation layer as described in Section 9.3.5. The SCCP as a signalling transport provides two different modes for RNSAP: connection-oriented and connectionless data transfer service.

RNSAP is responsible for bearer management signalling across the Iur interface. RNSAP is used for setting up radio links and allowing the SRNC to control those radio links using dedicated resources in a DRNS.

The RNSAP protocol uses one signalling connection per DRNC and UE where a UE may have one or more active radio links for the transfer of layer 3 messages.

The information transferred over the Iur interface can be categorised as follows:

- Radio and mobility control signalling: the Iur interface provides the capability to support radio interface mobility between RNSs for the UEs having a connection with UTRAN. This capability includes the support of handover, radio resource control and synchronisation between RNSs.
- Iub/Iur DCH data streams: the Iur interface provides the means for transport of uplink and downlink Iub/Iur DCH frames carrying user data and control information between SRNC and remote BS, via the DRNC.
- Iur DSCH data streams: an Iur DSCH data stream corresponds to the data carried on one

DSCH transport channel for one UE. A UE may have multiple Iur DSCH data streams. In addition, the interface provides a means to the SRNC for queue reporting and for the DRNC to allocate capacity to the SRNC.

- Iur RACH/CPCH data streams: in WCDMA FDD variant only.
- Iur FACH data streams.

The main UTRAN control functions supported by the RNSAP protocol exchange are:

- Transport network management.
- Traffic management of common transport channels, e.g. paging.
- Traffic management of dedicated transport channels, which includes, e.g. radio link set-up, addition and deletion as well as measurement reporting.
- Traffic management of downlink shared transport channels, which includes, e.g. radio link set-up, addition and deletion as well as capacity allocation.
- Measurement reporting for common and dedicated measurement objects.
- SRNS relocation, this function co-ordinates the activities when the serving RNS role is to be taken over by another RNS.

The RNSAP protocol participates in the radio link set-up procedure when it is used for establishing the necessary resources in the DRNC for one or more radio links. The connection-oriented service of the signalling bearer shall be established in conjunction with this procedure.

Other UTRAN-wide procedures, which may require protocol activities from the RNSAP entities are soft and hard handovers, cell update, URA update and paging as well as downlink power control actions and the RRC connection release and re-establishment whenever it needs to be performed over the Iur interface.

The complete specification of the RNSAP protocol is in the 3GPP specification TS 25.423.

9.4.1.3. NBAP – Node B Application Protocol

NBAP is a radio network layer protocol, which maintains control plane signalling across the Iub interface and thus controls resources in the Iub interface and provides a means for BS and RNC to communicate (Figure 9.29). The 3GPP specification TS 25.433 defines functionality of the NBAP protocol. One peer entity of NBAP resides in BS and for each BS the other entity resides in that RNC, which controls the BS, i.e. in the CRNC.

NBAP resides on top of the Iub transport layers and uses their services to transfer NBAP messages over Iub interface. The 3GPP R99 choice of underlying transport technology used to carry NBAP signalling over the Iub interface is ATM. The ATM connection together with the necessary ATM adaptation layers provides a signalling bearer for NBAP as defined in the 3GPP specification TS 25.432. The signalling bearer for NBAP provides reliable point-to-point connection. This means that NBAP assumes that each message NBAP sends to its peer entity will reach its destination without any errors. There may be multiple point-to-point links for NBAP signalling at the Iub interface. Since the Iub interface conforms to the generic model for UTRAN terrestrial interfaces, the specification of NBAP is also done so that functionality of NBAP does not depend on the transport layers below. New transport layers can be used with NBAP without the need to redefine the functionality of the protocol itself.

Control Plane Protocols in Radio Network:

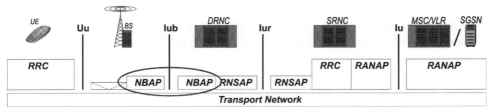

User Plane Protocols in Radio Network (PS Domain only):

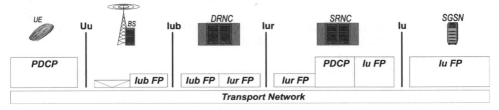

Figure 9.29.   Radio network – NBAP

BS does not make any decision of its own about the radio resource or RAB management but rather obeys commands from the RNC. NBAP offers its services to the control units of RNC and BS. NBAP also has an interface with the RRC protocol to provide the RRC entity in BS with information, which should be mapped to the BCCH logical channel. NBAP has no other interfaces to radio network layer protocols of UTRAN and all interactions with them are carried out via the control unit of RNC.

Figure 9.30 illustrates the operational environment of the NBAP protocol and all the interfaces of NBAP are depicted as bold black arrows on the figure.

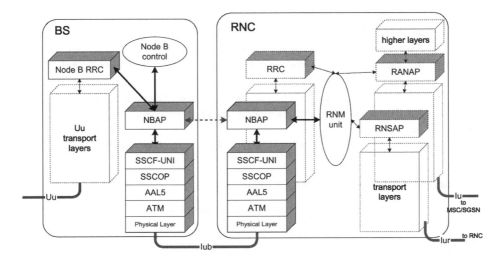

Figure 9.30.   Operational environment of NBAP

All the resources physically implemented in BS logically belong to and are controlled by RNC. Therefore, the physical resources of a BS are seen as logical resources by the CRNC. This RNC performs the control on the level of logical resources, while the physical implementation of resources and control procedures in the BS is left unspecified. Physical procedures performed in the BS change the conditions of the logical resources owned by RNC and this requires some information exchange between RNC and the BS to keep the RNC aware of the changes. Such information exchange is referred as Logical Operation and Maintenance (Logical O&M) of the BS. Logical O&M constitutes an integral part of NBAP signalling.

Another important responsibility of NBAP is to establish and maintain a control plane connection over the Iub interface, to initiate set-up and release of dedicated user plane connections across the Iub interface and command the BS to activate resources for new radio links over the Uu interface. All NBAP signalling functions are divided into common procedures and dedicated procedures.

The common NBAP procedures are not related to any particular UE but to common resources across the Iub interface. Signalling related to Logical O&M of the BS constitutes the great part of NBAP common procedures. Common NBAP includes procedures for configuration management of logical resources and procedures, which enable the BS to inform the RNC about the status of logical resources in the BS. Common NBAP also allows the RNC to initiate specific measurements in the BS and the latter to report the results of the measurements. Besides Logical O&M, common NBAP is used to deliver from RNC to BS the information to be transported on broadcast channel. Common NBAP also initiates the creation of a new UE context for each specific UE in BS by setting up the first radio link for that UE. Common NBAP maintains a control plane signalling connection across the Iub interface. For this reason there is always one signalling link for common NBAP at the Iub interface, which terminates at the common control port of the BS.

The dedicated NBAP procedures are always related to one specific UE, for which UE context already exists in BS. Dedicated NBAP includes procedures for management and supervision of existing radio links, whose UE is connected to the BS. These procedures allow the RNC to command the BS to establish or release some radio links for the UE context. Dedicated NBAP also provides the BS with ability to report failure or restoration of transmission on radio links. Dedicated procedures also include radio link reconfiguration management and support measurements on dedicated resources and corresponding power control activities, which allow the RNC to adjust downlink power level on the radio links. Dedicated NBAP signalling is carried out via one of the communication control ports of the BS logical model. There can be several dedicated signalling links for dedicated NBAP across the Iub interface.

The complete specification of the RNSAP protocol is in the 3GPP specification TS 25.433.

9.4.1.4. Radio Resource Control (RRC) Protocol

The RRC protocol is the key radio resource control protocol within UTRAN (Figure 9.31). It supports the RRM functionality discussed in Chapter 4 by coordinating the execution of resource control requests, which result from the decisions made by the RRM algorithms. With the help of the RRC protocol the effect of radio resource management decisions can be extended to all the UTRAN network elements affected by such decisions. The RRC protocol

Control Plane Protocols in Radio Network:

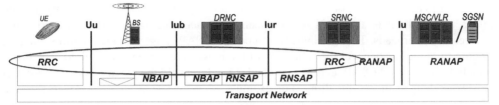

User Plane Protocols in Radio Network (PS Domain only):

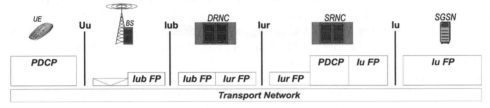

Figure 9.31. Radio network - (RRC) protocol

messages also carry in their payload all the signalling belonging to the non-access stratum protocols.

The RRC protocol operates between the UE and RNC. RRC protocol entities use the signalling bearers provided by the RLC sublayer to transport the signalling messages. All three modes of the RLC are available as different kinds of SAPs as shown in Figure 9.32. The

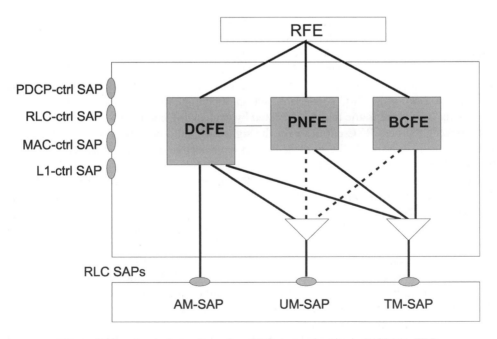

Figure 9.32. Logical structure of an RRC protocol entity in WCDMA-FDD

transparent mode TM-SAP is used, for example, whenever a UE has to communicate with an RNC prior to establishment of a full RRC connection (e.g. initial access or cell/URA update with a new cell). It is also used for frequently repeated messages, like paging, to avoid unnecessary overheads. The acknowledged mode AM-SAP is used for the control signalling specific to one UE whenever the reliability of the message exchange is required. The unacknowledged mode UM-SAP on the other hand is used to avoid the potential delay present in the acknowledged mode signalling. For example when releasing an RRC connection the release message is repeated many times (quick repeat function) via UM-SAP to increase the probability of the UE receiving it.

Figure 9.32 also shows the logical structure of an RRC protocol entity. The Dedicated Control Function Entity (DCFE) is used to handle all signalling specific to one UE. In SRNC there is one DCFE entity for each UE having an RRC connection with this RNC. The Paging and Notification Control Entity (PNFE) handles paging messages to the idle mode UEs. In a CRNC there is at least one PNFE entity for each cell to be controlled. The Broadcast Control Function Entity (BCFE) handles system information broadcasting on BCCH and FACH logical channels. There is at least one BCFE entity for each cell in a CRNC. Besides these functional entities a special routing function entity (RFE) is also modelled on top of Figure 9.32. Its task is to route the non-access stratum messages to different MM/CM entities on the UE side and to different CN domains on the RNC side.

As the key executor of the radio resource allocation the RRC entities in the UE and RNC have control over the L1, MAC and RLC entities on their side. In order to execute the control commands on these lower layer protocol entities special control SAPs are available to the RRC protocol entity as shown in Figure 9.32. These control SAPs are also used for reporting measurements and exceptional conditions detected by the lower layers.

The major function of the RRC protocol is to control the radio bearers, transport channels and physical channels. This is done by set-up, reconfiguration and release of different kinds of radio bearers. Before such actions can take place the RRC protocol communication itself must be initiated by using a number of (minimum four) signalling radio bearers. The resulting signalling connection together with any subsequently established other bearers is called *RRC connection*. During the RRC connection, set-up, reconfiguration and release of other radio bearers for user plane traffic may be executed by exchanging commands and status information between the peer RRC entities over the signalling radio bearers. The RRC connection will continue to exist until all user plane bearers have been released and the RRC connection between UE and RNC is explicitly released.

The security mechanisms applied on the radio bearers are activated and deactivated under the control of the RRC protocol. Besides the activation of confidentiality protection by ciphering, the RRC protocol can also guarantee the integrity of all higher layer signalling messages as well as most of its own signalling messages. This property, which is discussed more thoroughly in Chapter 8, is achieved by attaching to every RRC PDU a 32 bit message authentication code, which is calculated and verified by the RRC entities themselves. It ensures that the receiving RRC entity can verify that signalling data has not been modified and that the data has really originated from the claimed peer entity. Also the operation to change the keys used by these functions is performed by the RRC protocol under the instructions coming originally from the CN.

The UE mobility management at the UTRAN level is controlled by RRC signalling. Such mobility management functions executed by the RRC protocol are cell update, URA update and active set update, which have been discussed in Chapter 4.

Control and reporting of radio measurements is taken care of by the RRC protocol. A total of seven different categories of measurements can be activated at the UE, among them also the measurements for UE positioning. Each measurement can be controlled and reported independently from the other measurements.

Besides the above-mentioned tasks the RRC protocol also controls the broadcasting of system information and paging of UEs and exchange parameters for power control purposes.

As a protocol RRC is fairly complex. Altogether it has some 40 different procedures and more than 60 different kinds of PDU.

The RRC protocol is specified in the 3GPP specification TS 25.331 and a more detailed description can be found in Holma and Toskala (2001).

9.4.2. Radio Network User Plane

9.4.2.1. PDCP – Packet Data Convergence Protocol

As the name suggests the Packet Data Convergence Protocol (PDCP) is designed to make the WCDMA radio protocols suitable for carrying the most common user-to-user packet data protocols, TCP/IP (Figure 9.33). Therefore PDCP protocol entities can be found on both sides of the WCDMA radio interface: at the RNC and UE. The key functionality of the PDCP protocol is to compress the headers of the payload protocols, which – if sent without compression – would waste the invaluable radio link capacity. The only IP header compression specified for the 3GPP R99 is RFC2507 defined by the IETF.

In the 3GPP radio interface protocol model PDCP belongs to the radio link layer (L2), the topmost sublayer of it is specially designed for user-plane radio bearers carrying packet data.

Control Plane Protocols in Radio Network:

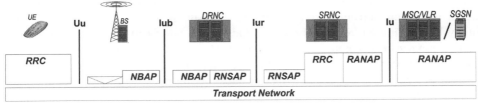

User Plane Protocols in Radio Network (PS Domain only):

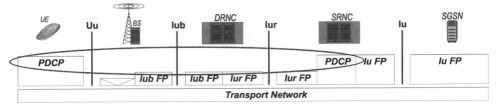

Figure 9.33.   Radio network user plane – PDCP

The PCDP protocol makes use of the RLC layer services in three different modes. Conceptually those services are available at three different Service Access Points (SAPs):

- Data transfer is acknowledged (AM-SAP)
- Data transfer is unacknowledged (UM-SAP)
- Data transfer in transparent mode (TM-SAP)

The PDCP protocol assumes that the segmentation and reassembly is taken care of by the RLC layer below. When lossless SRNS relocation is supported, then PDCP uses an AM-SAP and assumes in-sequence delivery of RLC SDUs.

PDCP uses the AM-SAP whenever reliable data transfer is requested. Then the RLC layer is only allowed to discard some of the RLC SDUs, if configured in such a way, but is expected to be able to indicate the PDCP of each the discarded SDUs both in the transmitter and the receiver. These indications are made necessary by the fact, that – as a compression protocol – the PDCP avoids sending explicit sequence numbers over the radio, but instead makes use of a virtual sequence numbering scheme. This makes it also necessary to let the PDCP know about possible resetting of the RLC connection and radio bearer reconfigurations, especially when lossless SRNS relocation is carried out.

For TM-SAP the size of the PDCP PDUs is expected to be fixed, which means very strict requirements either for the PDU size of the upper layer (e.g. IP protocol) or for the applied IP header compression. In a generic case neither the upper layers in the PS domain nor the currently specified IP header compression method can satisfy these requirements and therefore TM-SAP should be used very carefully and in well studied cases only.

The services provided by the PDCP protocol to the layers above are simply as follows:

- Delivery of packets (SDUs) either in acknowledged, unacknowledged or transparent mode.
- Radio bearer control with set-up, reconfiguration and release of a radio bearer in the PS domain.

The PDCP protocol entity running at the RNC has to interwork with the UTRAN frame protocols in order to relay the data packets in the downlink direction from Iu to Iub/Iur and in the uplink direction from Iub/Iur towards Iu. In normal cases this is achieved simply by copying packets from one interface to the other and performing the specified header compression/decompression on the way. Things get more complicated whenever an intersystem handover or lossless SRNS relocation needs to be performed. The PDCP protocol is therefore equipped with special interworking rules with the Iu frame protocols and the RRC protocol which ensure that packets are not lost or duplicated and that the packet sequence numbering can be restored after the handover or relocation at the target system.

In the 3GPP release 4/5 another header compression algorithm called Robust Header Compression (ROHC) as defined in the IETF RFC3095 will be added to the PDCP specification.

The complete specification of the PDCP protocol is in the 3GPP specification TS 25.323 and more information can also be found in Holma and Toskala (2001).

9.4.2.2. UTRAN Frame Protocols

The UTRAN frame protocols are the radio network layer user plane protocols, which carry the UMTS user data over the common UTRAN transport network (Figure 9.34). They are active across the UTRAN interfaces Iu, Iur and Iub. They use the transport network service, which is somewhat interface-specific.

Control Plane Protocols in Radio Network:

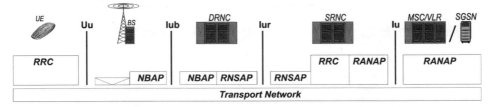

User Plane Protocols in Radio Network (PS Domain only):

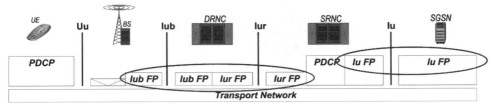

Figure 9.34.   Radio network user plane – UTRAN frame protocols

Frame Protocol at Iu

The frame handling protocol in the Iu interface has been designed to meet the needs of both the CS domain and PS domain user data traffic. The transport services for these two domains are different and therefore two modes of Iu frame protocol are used:

Transparent mode, which is used on such radio access bearers, which do not require any special protocol activities, like framing. Therefore the user data in this mode is simply passed through to the transport network stack without adding any protocol header information. Transparent mode hence becomes a null protocol. It is used, e.g. for carrying non-real time data packets in their plain GTP-U format from RNC to SGSN and vice versa.

Support mode, which takes the form of a real protocol with control PDUs and framing of user data. These protocol formats are applied on radio access bearers, which carry, e.g. AMR coded speech data in the CS domain. Therefore the support mode control PDUs can take care of rate control and time alignment, which are needed to support real-time speech transport. In this case the support mode protocol uses directly the AAL2 service provided by the underlying transport network.

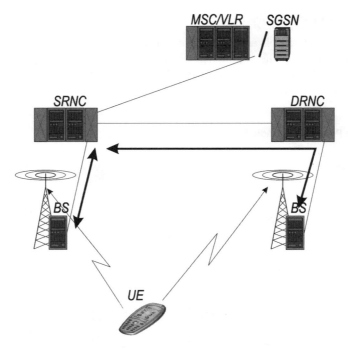

Figure 9.35.   Frame protocol for Iub/Iur dedicated channels

Frame Protocol for Dedicated Channels at Iur and Iub

There is a common frame protocol for dedicated channels over the Iub and Iur as illustrated in
Figure 9.35. With this protocol the SRNC can exchange user data frames with UEs being

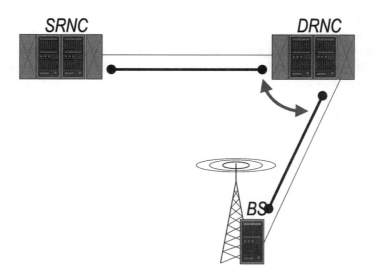

Figure 9.36.   Frame protocols for Iub/Iub common transport channels

serviced by its own as well as remote BSs. Besides the normal uplink and downlink data transfer, this frame protocol is capable of performing many control functions. Such protocol functions deal with timing adjustment and synchronisation. Also outer loop power control commands can be sent to UEs in dedicated channel frame protocol messages as well as other updates to the radio interface parameters, which need to be made known to the UEs.

Frame Protocols for Common Transport Channels at Iur and Iub

For sending and receiving data frames over common transport channels there are two different frame handling protocols: one for the Iub and another one for the Iur. Therefore, interworking between these protocols is done at the DRNC as illustrated in Figure 9.36.

Besides the uplink and downlink data transfer, which is to some extent specific to different transport channel types, these frame protocols also carry out control tasks. On the Iur frame handling flow control based on credits is applied. In the Iub frame handling synchronisation and timing adjustment needs are taken care of.

9.5. System Network Protocols

Once the radio network protocols make it possible to communicate across the UTRAN subnetwork by maintaining the communication path to the mobile terminals, it is the system network protocols, which create the communication services to the users of those terminals. The system (network) protocols operate on top of the radio network and within the UMTS CN itself.

9.5.1. Non-Access Stratum Protocols

The non-access stratum refers to the group of control plane protocols, which control the communication between UEs and CN. The common name "non-access stratum" refers to the fact that these protocols are carried transparently through the radio access network.

The protocols in this group belong to two sublayers of the system network as shown in Figure 9.3. The MM sublayer operates on top of the signalling connection provided by the radio network. Besides its own functionality, the MM sublayer acts as a carrier to the CM sublayer protocols.

9.5.1.1. MM Protocol for the CS Domain

MM protocol operates between UE and MSC/VLR (Figure 9.37). This control plane protocol provides basic signalling mechanisms for controlling mobility management and authentication functions between UEs and the CN CS domain. The MM protocol is supported by UEs operating either in the PS/CS operation mode or in the CS-only operation mode.

The MM protocol makes use of the signalling connection provided by the radio network layer protocols. This signalling connection is sometimes referred to as the radio resource (RR) connection, which consists of the RRC protocol connection over the signalling radio bearer and the RANAP connection over the Iu signalling bearer. Before any MM protocol activity can take place the signalling connection must first be established. Inside the CN the MM protocol entity on the MSC/VLR side has to interwork with the MAP protocol in order to maintain the location registration in the HLR.

Control Plane Protocols in System Network (CS Domain):

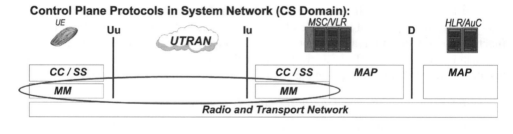

Control Plane Protocols in System Network (PS Domain):

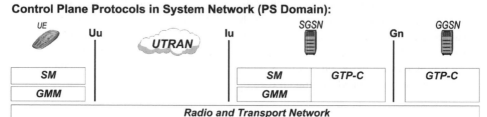

Figure 9.37.   System network non-access stratum – MM protocol for the CS domain

The MM protocol controls the MM state of the UE and the corresponding peer entity on the network side. The major states of these protocol entities indicate the location update status of the UE, which has a major impact on its ability to use the services of the CM sublayer. This is also why the MM protocol is used for carrying messages of the CM sublayer protocols between UE and the MSC/VLR.

The MM protocol has three basic categories of procedures:

- MM connection procedures
- MM common procedures
- MM specific procedures

The *MM connection procedures* are used to establish and release MM connections as well as for transferring the CM sublayer messages. The concept of MM connection is established without exchange of any special MM protocol initiation messages. Instead, an MM connection is established and released implicitly on request from a CM entity.

When the MM connection establishment is initiated by the UE, the MM protocol entity issues a CM SERVICE REQUEST, which contains the following parameters

- Identity of the UE, e.g. TMSI;
- Mobile station classmark 2;
- Ciphering key sequence number; and
- CM service type identifying the requested type of transaction, which can be, e.g. mobile originating call establishment, emergency call establishment, short message service or supplementary service activation.

The MM protocol entity receiving this request on the CN side will analyse these parameters and decide on approval and continuation of the service request.

On the other hand the request to establish an MM connection may come from the CM entity on the network side. This is then accomplished by making the UE to establish MM connection via the paging procedure.

After the MM connection has been established, it can be used by the CM sublayer entity for information transfer. According to the protocol architecture shown in Figure 9.3, each CM entity will have its own MM connection. These different MM connections are identified by the Protocol Discriminator (PD) and, additionally, by the Transaction Identifier (TI).

An established MM connection can be released by the local CM entity. The release of the connection will then be done locally in the MM sublayer, i.e. no MM messages are sent over the radio interface for this purpose.

The *MM common procedures* may be initiated at any time while MM connections are active. These procedures are used mainly for security functions when the identity and location of the UE is known by the CN and the MM connection exists from the CN to the UE. The mandatory MM common procedures are:

- *TMSI reallocation procedure*, which is used to provide confidentiality of the user identity, i.e. to protect a user against being identified and located by an intruder. In this procedure a temporary random identification (TMSI) is allocated by the MSC/VLR for use within radio interface procedures instead of the permanent, and therefore traceable, IMSI identification. The procedure is initiated by the CN and is usually performed in the ciphered mode.
- *Authentication procedure*, which permits the CN to check whether the identity provided by the UE is acceptable or not, and provide parameters enabling the USIM to calculate new ciphering and integrity keys. In a UMTS authentication challenge, the authentication procedure is extended to allow the UE to check the authenticity of the CN (see Chapter 8). This allows, for instance, detection of a false base station. The authentication procedure is always initiated by the CN.
- *Identification procedure*, which is used by the CN to request the UE to provide specific identification parameters to the network like IMSI and IMEI.
- *IMSI detach procedure* is initiated by the UE when the UE is deactivated or if the USIM is detached from the ME. Each UMTS network can broadcast a flag within the system information block to indicate whether the detach procedure is required by UEs. When receiving an MM IMSI DETACH INDICATION message, the network may set an inactive indication for this IMSI. No response is returned to the UE. After this the network shall release locally any ongoing MM connections, and start the normal signalling connection release procedure.
- *Abort procedure* can be used by the CN to abort any on-going MM connection establishment or already established MM connection due to a network failure or having found an illegal ME.

The *MM specific procedures* handle messages used in location update procedures. An example message flow of a location update procedure is described in Chapter 10. There are three variants of the basic location update procedure:

- *Normal location updating procedure* for the situation where the UE has detected that its location area is changed and it is necessary to inform the CN about the new location. The detection is done by checking the Location Area Identifier (LAI) value broadcasted by the network against the current LAI value stored in the USIM. The normal location updating procedure shall also be started if the network indicates that the UE is unknown in the VLR as a response to service request. The UE has to maintain a list of "forbidden location areas for roaming" on the USIM based on failed location update requests.

- *Periodic location updating procedure* may be used to notify periodically the availability and location of the UE to the CN according to a timer in the UE. The timeout value for this timer is taken from the broadcasted system information block.
- *IMSI attach procedure* is performed when the UE enters (or is switched on in) the network in the same location area, where it was previously detached, i.e. when the LAI value broadcasted in the current cell is detected to be equal to the LAI value on the USIM.

The common characteristic feature of the MM specific procedures is that they can only be initiated when no other MM specific procedure is running or when no MM connection exists. Notice also that any of the MM common procedures described above (except IMSI detach) may be initiated during a MM specific procedure.

Since the MM protocol is a direct extension of its GSM predecessor, a reader interested in more details about it may consult any of the GSM textbooks listed in the Bibliography. The complete specification of the MM protocol together with the UMTS specific details is found in 3GPP specification TS 24.008.

9.5.1.2. GPRS Mobility Management for PS Domain (GMM)

GMM operates between the UE and SGSN (Figure 9.38). The name GPRS MM is used because the protocol is basically evolved from the corresponding protocol for packet switched services in the GSM/GPRS system. This control plane protocol provides basic signalling mechanisms for controlling MM and authentication functions between UEs and the CN PS domain. The GMM protocol is supported by UEs operating either in the PS/CS operation mode or in the PS operation mode.

Like the MM protocol, the GMM protocol makes use of a signalling connection between

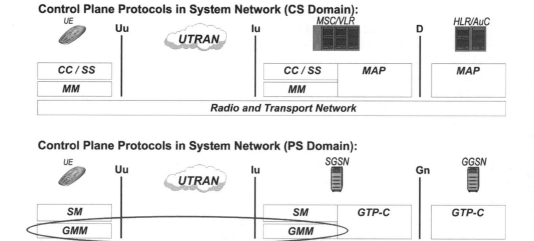

Figure 9.38.   System network non-access stratum – MM protocol for the PS domain

UE and SGSN. This signalling connection – which is sometimes called PS signalling connection – is provided by the radio network layer protocols using RRC over the signalling radio bearer and RANAP over the Iu signalling bearer. The GMM protocol entity on the SGSN side has to interwork with the GTP-C protocol entity for location information exchange.

It is possible that a UE operating in PS/CS operation mode has signalling connections both to the MSC/VLR and to SGSN simultaneously. For optimisation reasons it is defined that in these circumstances the GMM protocol entities in both UE and SGSN may perform so called combined procedures. The idea of the combined GMM procedures is to avoid sending both MM and GMM messages when the CN PS domain can inform the CN CS domain about such combined activities inside the CN. The optional Gs interface is used to take this action on the CN side.

From the packet data services point of view it is important to distinguish between the state, when the UE together with its identity and location is known by the CN PS domain and the state, when it is not known. The importance comes from the fact that the packet user data is protected by encryption which makes use of user-specific keys. This separation is highlighted by the *GMM context* concept. When the user is known to the CN PS domain network the GMM context exists, otherwise it does not exist.

GMM procedures for explicit establishment and release of the GMM context are:

- *GPRS attach* and *combined GPRS attach* procedures are used to indicate the user identity (typically P-TMSI) and current routing area to the network and to establish a GMM context for this user. A typical situation to execute this procedure is when the packet mode UE is switched on or USIM is inserted into it. In the case of the combined GPRS attach procedure the activities for normal IMSI attach are carried out.
- *GPRS detach* or *combined GPRS detach* procedures are invoked when the UE is switched off or the USIM is removed from the UE or a network failure has occurred. The procedures cause the UE to be marked as inactive in the CN PS domain only or in the combined GPRS detach procedure also for the CN CS domain.

The following GMM procedures are used for location management when a GMM context already exists:

- Normal *routing area updating* and *combined routing area updating* procedures are performed when the UE recognises that the routing area is changed to inform the CN PS domain about the event. In the combined routing area updating the location area is also included for the CN CS domain.
- *Periodic routing area updating* procedures may be used to notify periodically the availability of the UE to the CN according to a timer in the UE.

The following GMM common procedures for security purposes can be initiated when GMM context for the user has been established.

- A *P-TMSI re-allocation* procedure provides a temporary random identification, packet-TMSI (P-TMSI) for use within radio interface procedures instead of the permanent, and therefore traceable, IMSI identification. Normally the P-TMSI re-allocation procedure is performed in conjunction with another GMM procedure like routing area updating.
- A *GPRS authentication and ciphering* procedure, initiated always by the CN, permits the CN to check whether the identity provided by the UE is acceptable. It also provides parameters enabling the USIM to calculate UMTS ciphering and integrity keys and

permits the UE to authenticate the CN. See Chapter 8 for more details about these security functions.

- A *GPRS identification procedure* is used by the CN PS domain to request the UE to provide specific identification parameters to the network like IMSI and IMEI.

One more GMM procedure is needed to provide services to the CM sublayer on top of the GMM protocol:

- A *service request* procedure, which is initiated by the UE, is used to establish a secure connection to the network and/or to request the bearer establishment for sending data. The procedure is typically used when the UE has such a signalling message (e.g. SM or paging response message) or user data packet, that requires security protection, to be sent.

More details about the GMM protocol can be found in the 3GPP specification TS 24.008.

9.5.1.3. Call Control (CC) Protocol

The CC protocol provides basic signalling mechanisms for establishing and releasing circuit switched services (e.g. voice call or multimedia call) (Figure 9.39). The CC protocol operates between an MSC/VLR and a UE operating in the CS/PS or CS-only mode. The CC protocol uses the connection service provided by the MM sublayer to carry the CC protocol messages.

Control Plane Protocols in System Network (CS Domain):

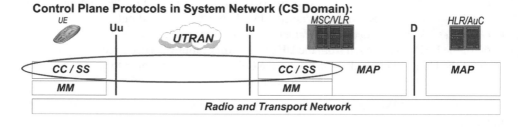

Control Plane Protocols in System Network (PS Domain):

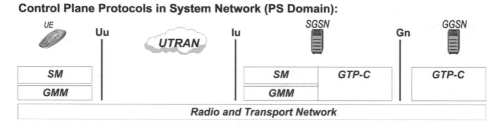

Figure 9.39. System network non-access stratum – CC protocol

The normal circuit switched voice call is controlled pretty much like in the ISDN telephony. Therefore the CC protocol entity in the GMSC has to interwork with the ISDN User Part (ISUP) protocol for establishment of a call path through an external ISDN network.

The UMTS circuit switched multimedia call is based on the 3G-324M. The 3G-324M is a variant of the ITU-T recommendation H.324. For details see the 3GPP specification TS 26.111.

The call establishment procedures establish the CC connection between the UE and the CN and activate the voice/multimedia codec and interworking functions, if needed. The CC protocol entity in the MSC/VLR also interworks with the RANAP protocol entity in order to establish a RAB for the circuit switched call. This means, that the QoS requirements received by the CC protocol entity in the CC SETUP message have to be mapped onto a RAB request for the RANAP protocol entity in the same MSC/VLR. Only after the UTRAN has confirmed the assignment of a RAB, the call establishment can then proceed.

The established calls can be Mobile Originating (MO) or Mobile Terminating (MT). For multimedia circuit switched calls the normal call establishment procedure applies with some extra information elements in the CC protocol messages. The call release procedure, initiated by the UE or the CN release the resources for the call.

Call collisions as such cannot occur at the network, because any simultaneous MO or MT calls are dealt with separately assigned and different transaction identifiers in the CC protocol.

CC as a core network function was discussed in Chapter 5. An example of message flows for circuit switched call establishment and release are given in Chapter 10.

Since the CC protocol is a direct extension of its GSM predecessor, a reader interested in more details about it may consult any of the GSM textbooks listed in the Bibliography. The full specification of the CC protocol together with the UMTS specific details can be found in the 3GPP specification TS 24.008.

9.5.1.4. Session Management (SM) Protocol

The protocol entities running the SM protocol are located in the UE and SGSN (Figure 9.40). This protocol is the counterpart to the circuit switched CC protocol in the sense, that it is used to establish and release packet data sessions. To underline the possibility of using several packet data protocols within the UMTS network these sessions are called PDP contexts as discussed in Chapter 5. In practice the single most important packet data protocol to be supported is IP and especially the IPv6.

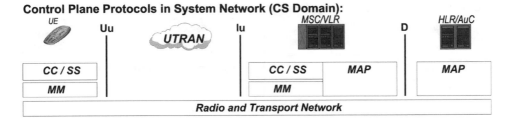

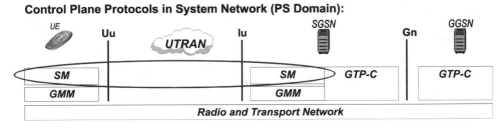

Figure 9.40.   System network non-access stratum – SM protocol

The main purpose of the SM protocol is to support PDP context handling of the UE. In SGSN the SM protocol entity interworks with the GTP-C protocol entity on the Gn interface side in order to extend the PDP context management to the GGSN. The PDP context contains the necessary information for routing user plane data packets from the GGSN to the UE and vice versa. A UE may have more than one PDP context activated simultaneously in which case each of them is controlled by a SM protocol entity of its own.

The SM procedures require the existence of a GMM context for the identified user. If no GMM context for the user has been established, the MM sublayer has to initiate a GMM procedure to establish the needed GMM context (see GMM protocol above). After the GMM context establishment, the SM protocol uses services offered by the MM sublayer. Ongoing SM procedures are suspended during a GMM procedure execution.

The SM procedures can be initiated either by the GGSN or by the UE. Basic SM procedures are:

- The PDP context activation procedure, which establishes a new PDP context at the UE, SSGN and GGSN with necessary routing and QoS parameters for the session.
- The PDP context modification procedure is invoked when a change in the session parameters like QoS or traffic flow template is needed during the session. The modification may be initiated either by the UE or the GGSN.
- The PDP context deactivation procedure destroys PDP contexts in the UE and CN elements when the corresponding session is no longer needed.

A more detailed description on the SM protocol is given in the 3GPP specification TS 24.008.

9.5.1.5. Supplementary Services (SS) Protocol

Supplementary services are additional services defined to be used over circuit switched connections (Figure 9.41). This concept has been inherited from the 2G GSM system and

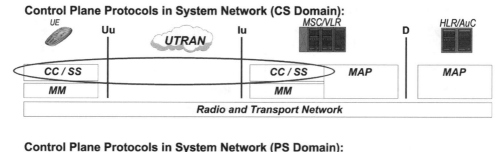

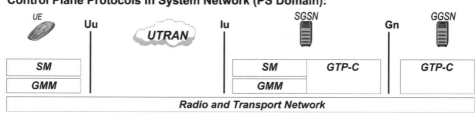

Figure 9.41. System network non-access stratum – SS protocol

therefore a UMTS terminal operating in the CS mode has the same ability for SS as a GSM mobile phone.

The 3GPP R99 implementation supports all the SS as far as circuit switched traffic is concerned. Some of the typical supplementary services are:

- Call forwarding services with different triggering conditions: no reply, busy, unconditional, not reachable.
- Call barring services for all incoming, all outgoing, all outgoing international or roaming calls.
- In-call services like call wait and call hold.

A complete list and description of SS, which are supported by the 3GPP R99 can be found in specification TS 24.080.

SS are controlled by the SS protocol, which is part of the CM sublayer. The SS protocol entities on the UE and MSC/VLR side communicate by using the MM connection. The communication has to do with activation and deactivation of the supplementary services. On the CN side the SS protocol interworks with the MAP-protocol, which is used whenever the supplementary service command requires, e.g. access to the subscriber information in the HLR.

9.5.2. Control Plane Between CN Nodes

Section 9.5.1 briefly introduced the control plane protocols in the non-access stratum between the UE and the CN. Despite the fact that the CN is divided into two functional domains, circuit and packet switched domains, these entities also have to perform some signalling functions with each other and especially with the registers located in the home network part of the CN.

MAP and CAP protocols have already been mentioned in Section 9.1.3 and an interested reader may find more detailed descriptions on these protocols in the GSM textbooks listed in the Bibliography. The 3G version of the MAP protocol is defined in the 3GPP specification TS 29.002.

9.5.2.1. GTP-C – GPRS Tunnelling Protocol for Control Plane

GTP-C belongs to the PS domain in the CN and is located in SGSN and GGSN (Figure 9.42). GTP-C is a control plane protocol specifying tunnel management and control procedures in order to allow SGSN and GGSN to provide user data packet transfer. GTP-C is also used to transfer MM signalling messages between SGSNs, which corresponds to the MAP/G interface on the circuit switched side.

GTP-C operates on two different interfaces, Gn and Gp. The Gn interface is shared between GSNs (i.e. SGSNs and GGSNs) belonging to the UMTS packet domain network. The Gp interface is used for UMTS network interworking between two UMTS packet domain networks. See Figure 5.4.

On both Gn and Gp interfaces GTP-C operates on top of the UDP/IP protocol family. UDP provides connectionless message transfer without the need to establish a connection. The main service of IP is message routing between source and destination network elements.

In SGSN the GTP-C protocol entity interworks with the GTP-U protocol entity at the Gn side and with the GMM and SM protocol entities also. In GGSN the GTP-C entity interworks

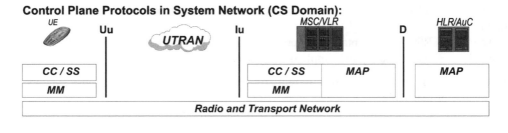

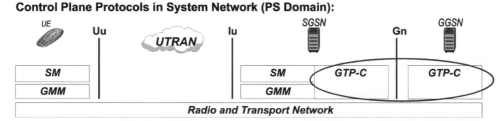

Figure 9.42. Control plane – GPRS tunnelling protocol for the control plane

with the GTP-U protocol entity located at the Gn interface and with the external network interworking function at the Gi interface, e.g. in order to allocate dynamic addresses for PDP contexts.

The GTP-C protocol is required to be able to transfer its signalling messages reliably over the Gn and Gp interfaces. Therefore the design of the GTP-C had to solve the problem of unreliable message transfer provided by the UDP/IP transport layers. Thus GTP-C includes timers and message sequence numbers needed to control and detect message loss and message retransmission.

GTP-C Procedures Between SGSN and GGSN

GTP-C is used to create, delete and modify GTP tunnels that are used to transfer user data packets. GTP-C also checks if the path towards the peer GTP-C is alive.

When a PDP context is activated, GTP-C in SGSN initiates GTP tunnel establishment towards GGSN. The purpose of the GTP tunnel establishment procedure is to negotiate parameters related to user data transfer between SGSN and GGSN. Such parameters are QoS and parameters associated with the activated address, like the traffic flow template for example.

When a PDP context is deactivated, GTP-C removes GTP tunnelling information from the SGSN and GGSN.

The PDP context modification procedure provides a flexible and dynamic way to react to changed conditions. When a PDP context is modified, e.g. due to changed QoS attributes or load condition, GTP-C initiates a GTP tunnel modification procedure. In the GTP tunnel modification procedure both the SGSN and GGSN are able to modify almost all the parameters negotiated in GTP tunnel establishment procedure. The PDP context modification procedure is also performed as part of inter-SGSN routing area update procedure and SRNS relocation procedure involving two SGSNs to notify GGSN of the new SGSN starting to serve the UE.

In the case of network requested PDP context activation is supported and GGSN does not include protocols needed to communicate with HLR across the Gc interface, GTP-C is used to transfer messages to another GSN that converts messages between GTP and MAP. MAP protocol is then used to communicate across the Gc interface towards HLR. The basic need for a GGSN to communicate with HLR is to find out if the UE is attached to the network and also the address of the SGSN serving the UE. This information is needed by the GGSN before it can request the SGSN to initiate a PDP context activation procedure.

GTP-C Procedures Between SGSNs

GTP-C is used in communication between two SGSNs in order to exchange information related to a UE. The communication is initiated when a UE makes a GPRS attach procedure, inter-SGSN routing area update procedure or UTRAN makes the decision to perform the SRNS relocation procedure involving two SGSNs.

The use of GTP-C in exchanging UE related information between SGSNs minimises the need to use a radio interface. For example when an inter-SGSN routing area update procedure or SRNS relocation procedure involving two SGSNs is performed, the new SGSN serving the UE receives almost all the needed information of the UE from the old SGSN. Such information consists of GMM related parameters, like IMSI and GMM context, and SM related parameters, like active PDP contexts.

The use of GTP-C in exchanging UE related information between SGSNs also strengthens the system security. For example when a UE makes a GPRS attach procedure, IMSI may be retrieved from the old SGSN and not from the UE over the radio interface.

The GTP-C protocol is defined in the 3GPP specification TS 29.060.

9.5.3. User Plane in the System Network

In the case of the CS domain the role of user plane protocols is taken by the speech codec

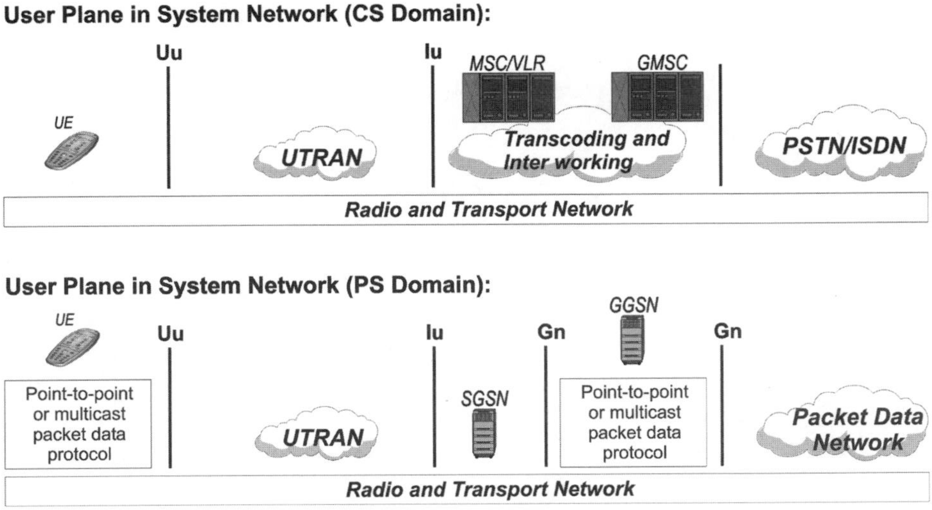

User Plane in System Network (CS Domain):

User Plane in System Network (PS Domain):

Figure 9.43.   User plane in the system network

functions, which are active in the UE and serving MSC/VLR (Figure 9.43). In 3GPP R99 the predefined speech codec is Adaptive Multi Rate (AMR) codec, which was discussed in Chapter 7. Besides transcoding the interworking functions, mentioned already in Chapter 5, have to be taken care of between the UMTS and external PSTN/ISDN network.

In the case of the PS domain the UE and GGSN exchange data packets according to some point-to-point or multicast packed data protocol. For the time being the most common user plane protocol will continue to be the IP protocol.

10

Procedure Examples

In order to allow the reader to accomplish a network-wide view of some of the most important UMTS functionalities discussed in the preceding chapters we present a few examples of UMTS system procedures. These procedures illustrate the co-ordination of actions carried out by all network entities and the role of the UMTS protocols in controlling this co-ordination.

10.1. Elementary Procedures

In this chapter a basic model for system-wide procedures is presented by modelling each system procedure as a kind of communication-intensive transaction. The concept of *transaction* is used in order to underline the fact that these system-wide procedures are executed into well-defined completion and that subsequent procedures represent fairly independent instances of communication.

The layering and interworking of protocols as described in Chapter 9 makes it possible to distinguish a set of elementary procedures, which can be used as building blocks in the design of network transactions. Depending on a transaction and its type – whether it is mobile terminated or mobile originated – some parameters and messages vary inside the elementary procedures and some elementary procedures may or may not be used.

The UMTS protocol interworking model (see Figure 9.4) is needed here in the sense that different procedures make use of different layers in the UMTS network. The transport network is used in every procedure: signalling transport is always required and user data transport is required by many transactions. Radio network functions are needed whenever access network services are required within an elementary procedure. The overall control of system transactions is done by the system network protocols, which invoke the elementary procedures in a stepwise manner and determine how the transaction is proceeding through different steps.

Basically any network transaction can be divided into eight steps as presented in Figure 10.1. For each step an elementary procedure can be distinguished.

Paging is a mobility management procedure used when searching certain subscriber from the network coverage area. This procedure is only executed if the transaction originates from the network side. The remaining seven steps are the same whether the transaction is originated or terminated by the UE.

Radio Resource Control (RRC) connection set-up is an elementary procedure containing activities and message flow to establish a radio control connection between the terminal and radio access network. The details of this procedure vary depending on the case.

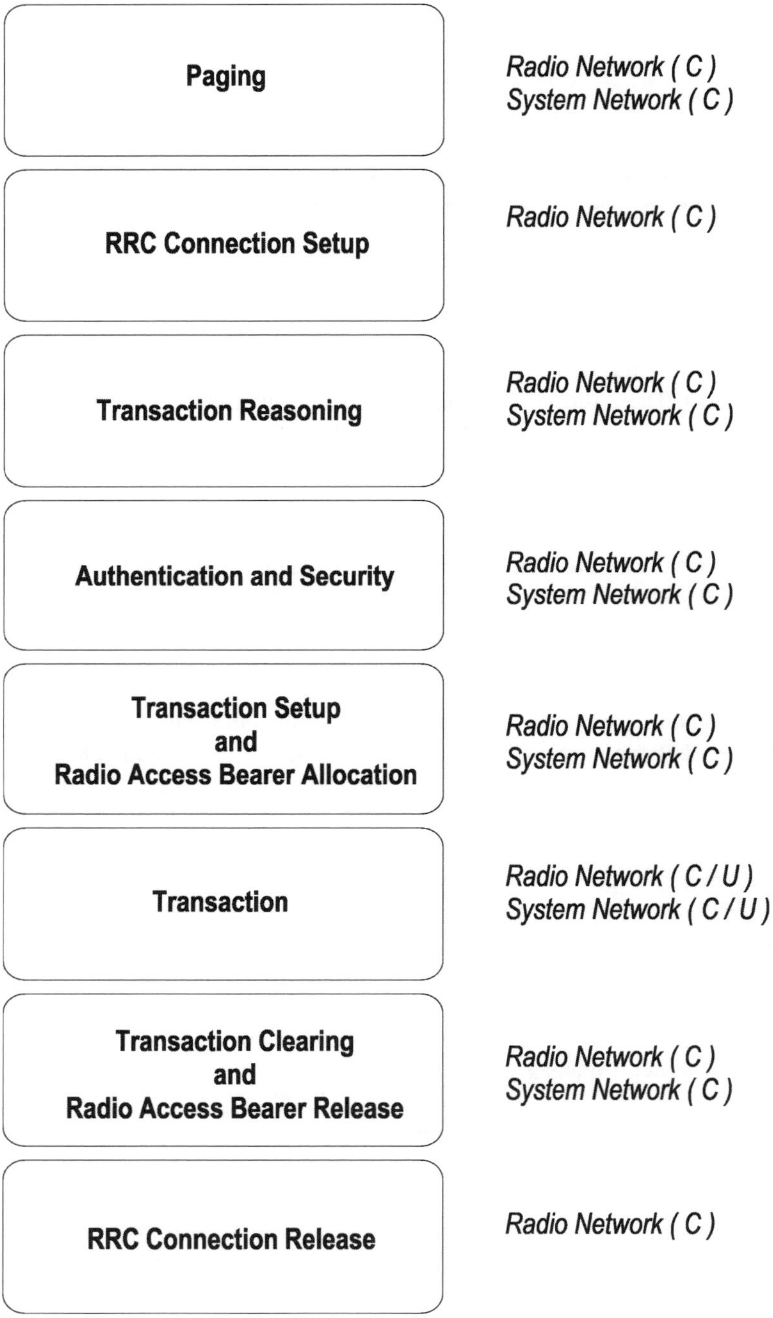

Paging	*Radio Network (C)* *System Network (C)*
RRC Connection Setup	*Radio Network (C)*
Transaction Reasoning	*Radio Network (C)* *System Network (C)*
Authentication and Security	*Radio Network (C)* *System Network (C)*
Transaction Setup **and** **Radio Access Bearer Allocation**	*Radio Network (C)* *System Network (C)*
Transaction	*Radio Network (C / U)* *System Network (C / U)*
Transaction Clearing **and** **Radio Access Bearer Release**	*Radio Network (C)* *System Network (C)*
RRC Connection Release	*Radio Network (C)*

Figure 10.1.   Basic model of UMTS network transactions

Transaction reasoning is an elementary procedure where the terminal indicates to the Core Network (CN) which kind of transaction is requested. Based on the reasoning information the CN may decide how to proceed with the transaction or decide to terminate the execution of the transaction.

After these steps the *authentication and security* procedure normally takes place. This elementary procedure authenticates the UMTS subscriber and network to each other and later activates the necessary security mechanisms for the access network connection.

Transaction set-up with radio access bearer allocation is an elementary procedure which allocates the actual communication resources for the transaction. Therefore the details of this procedure depend on the type of the transaction (circuit or packet switched).

The elementary procedure *transaction* is the phase where the user plane connection exists, i.e. the UE has a UMTS bearer connection active across the whole UMTS network.

Transaction clearing with radio access bearer release is an elementary procedure used for releasing the network resources related to the transaction.

The *RRC connection release* is an elementary procedure containing mechanisms with which the radio control connection between the UE and the access network is released.

10.1.1. Paging

The paging procedure is needed whenever a UE terminating transaction is to be performed. Unlike in GSM, UMTS contains two types of paging methods, Type 1 and Type 2.

Paging Type 1 is used by the CN domains and this is the "conventional" way to use paging (see Figure 10.2). Paging is part of the radio network control plane and it is delivered over the Iu Interface by using the *RANAP Paging* message. The CN domain initiating paging addresses it to those LAs/RAs, in which the desired UE has last reported its location, i.e. performed either location update or routing area update.

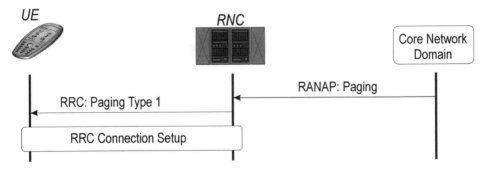

Figure 10.2.   Paging Type 1

An RANAP paging message contains two mandatory parameters, the requesting CN domain and IMSI. From a system security point of view, the unnecessary transfer of IMSI over the Uu reference point without encryption is not desired and because of this the RNC may or may not perform various conversions for the IMSI. For example, the RNC may issue a *RRC Paging Type 1* message containing Radio Network Temporary Identity (RNTI) instead of IMSI. This value has been assigned when the UE is attached to the network or performed the previous transaction.

Typically the receipt and recognition of the RRC paging Type 1 message by the UE leads to the establishment of an RRC connection, which is needed to carry on the transaction, which was initiated by the paging.

In addition to the previously described use, the RRC paging Type 1 has some RAN specific uses. When the UE is not in idle mode and changes in the system information parameters need to be made known to the UE, the RNC may use RRC paging Type 1 to "awake" the UE and thus make it able to read the revised system information or certain parts of it.

If the UE has a packet switched connection without any activity the RNC uses RRC paging Type 1 to inform the UE that the packet switched connection should be reactivated. In this case there are some packet data the CN desires to deliver to the UE. This forces the UE to perform cell update and after this the RNC is able to route the data packets to the correct UE.

UMTS terminals are meant to be able to handle various connections simultaneously. This is, they may have more than one connection at a time with CN domains. When the UE already has a connection with one CN domain and another connection is established with the same domain the CN sends RANAP paging like in previous cases. The paging sent to the UE by the SRNC is now *RRC paging Type 2* (see Figure 10.3).

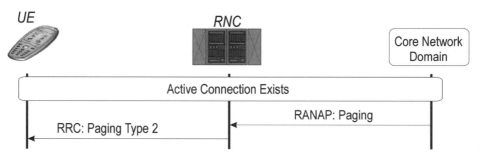

Figure 10.3.   Paging Type 2

The separation between paging Type 1 and Type 2 is done in the RNC, the RANAP paging always looks the same. The difference between Type 1 and Type 2 is that Type 1 is used for paging those UEs which are in idle mode, cell-PCH or URA-PCH state and hence it could be either for certain UEs or for all UEs within cell(s). Type 2, on the other hand is used for paging those UEs, which are in cell-DCH or Cell-FACH state, therefore it is always dedicated and addressed to one UE only.

10.1.2. RRC Connection Set-up

Figure 10.4 illustrates the principle of how the radio connection between the UE and the RNC is established over the Uu interface and the access domain internal interface Iub.

The RRC connection set-up always starts from the UE with the message *RRC Connection request* sent over CCCH (Common Control Channel). Referring to Chapter 3, the CCCH in the uplink direction means RACH (random access) and actually the RRC connection request content is coming from the UE over physical RACH as a result of the initial access procedure. This message arrives to the RNC over the Iub RACH data port. Upon receiving this message the RRC entity at the RNC side changes its state from IDLE to CONNECTED Cell_FACH or

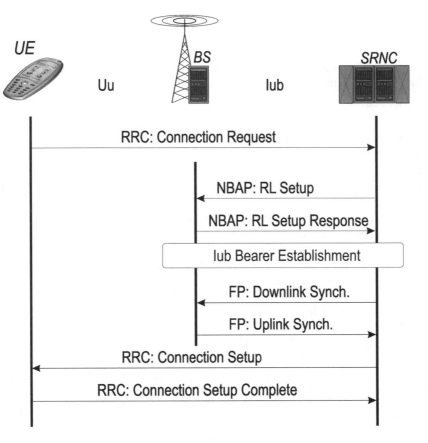

Figure 10.4. RRC connection set-up

Cell_DCH. Then the RNC communicates towards the UE over common control channels and from those, the used ones are FACH and RACH, respectively.

The RRC connection request message contains plenty of information related to the requested radio connection, terminal and subscriber identity; in this message the UE sends to the network, for instance, IMSI or TMSI, IMEI, location area identity and routing area identity. The RRC connection request must indicate which of these values are inserted in the message and also contain those values accordingly. In addition to these identities the RRC connection request contains the reason why the radio connection is requested. There are various reasons and the following list shows some of them:

- Originating conversational call
- Originating streaming call
- Originating interactive call
- Originating background call
- Terminating conversational call
- Terminating streaming call
- Terminating interactive call

- Terminating background call
- Emergency call
- High priority signalling
- Low priority signalling
- Call re-establishment

As can be seen from the list, the RRC connection request already indicates what kind of QoS will be requested when the transaction proceeds. Emergency call is separated because it will be treated differently over the network than normal transactions. If the subsequent transaction will be only signalling, this is indicated, too.

Depending on the request reason, the RNC makes a decision whether to allocate dedicated or common resources to this transaction. Based on this decision-making, the SRNC allocates RNTI and other resources to this transaction.

The Iub interface is "opened" when the RNC sends a *NBAP radio link set-up* message to the BS. This message contains transport format description, power control information and code information, which is the uplink scrambling code for WCDMA-FDD. The BS acknowledges this message by sending a *NBAP radio link set-up response* message. This message informs the RNC about transport layer addressing information (AAL2 address) and gives also some reference information for establishment of an Iub bearer in the transport network.

The serving RNC starts the Iub bearer establishment according to the information received from the BS. This procedure is carried out by the internal control plane of the transport network at the Iub interface. The established Iub bearer is bound together with the DCH (Dedicated Channel) assigned to the transaction. After this the frame protocol connection over Iub is synchronised with message exchange. If we suppose that a dedicated radio connection using DCH is set up here, then it is the dedicated channel frame protocol for Iub.

When the Iub communication is ready the RNC sends the *RRC connection set-up* message to the UE over common control channels (FACH in this case). In this message the SRNC informs the UE about the transport format, power control and codes, which in the case of WCDMA-FDD is a downlink scrambling code. The UE confirms RRC connection establishment by sending the message *RRC connection set-up complete*.

10.1.3. Transaction Reasoning

When the RRC connection has been established the UE sends the message *RRC initial direct transfer*. RRC initial direct transfer carries in its payload the first system network message of the transaction from UE to the network (Figure 10.5).

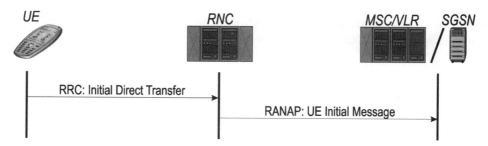

Figure 10.5.   Transaction reasoning

Upon receiving this information the RNC adds some more parameters to it and forwards this combination to the appropriate CN domain as a *RANAP UE initial message*. The RANAP UE initial message contains in its payload the contents of the original RRC direct transfer message together with the first system network message generated by the UE.

The contents of the UE initial message provide the CN with lot of information about the transaction initiated by the UE. Among such information is the (claimed) identity of the UMTS subscriber (TMSI or IMSI), his/her current location area and the kind of transaction requested. All this information is used by the CN node (MSC/VLR or SGSN) to decide, how to proceed with the transaction request.

10.1.4. Authentication and Security Control

In Chapter 8 we discussed security functions and algorithms. Figure 10.6 shows the message flow related to access network security in the context of a transaction.

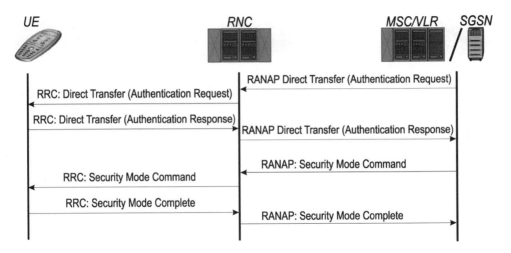

Figure 10.6.   Authentication and security control

During the RRC connection establishment the UE already informed the RNC with the UE classmark parameter about its capabilities in many respects, one of those being the security algorithms supported by the UE.

The UE and the network authenticate each other by sending the *authentication request* message in the payload of RANAP and RRC direct transfer messages to the UE. After executing the authentication algorithms on USIM the UE responds with an *authentication response* message again in the payload of RRC and RANAP direct transfer messages. In this dialogue the RNC acts as a relay forwarding the contents of RANAP direct transfer to RRC direct transfer and vice versa.

With a *RANAP security mode command* message the related CN domain indicates to UTRAN that the transaction should be encrypted. This message indicates the selected security algorithms and delivers the integrity and encryption keys to UTRAN.

Based on this information the RNC commands the UE to start encryption with the corresponding keys and algorithms by sending the *RRC security mode command*. By issuing an *RRC security mode complete* message the UE indicates that it has successfully turned on the selected integrity protection algorithm and encryption algorithm for protection of further communication in this transaction.

The RNC still has to inform the related CN domain about the procedure completion with the message *RANAP security mode complete*.

10.1.5. Transaction Set-up with Radio Access Bearer (RAB) Allocation

So far, the contents of the elementary procedures have been similar and thus independent on the CN domains. Transaction set-up with radio access bearer allocation is the first step where the different nature of CN domains has to be taken into account.

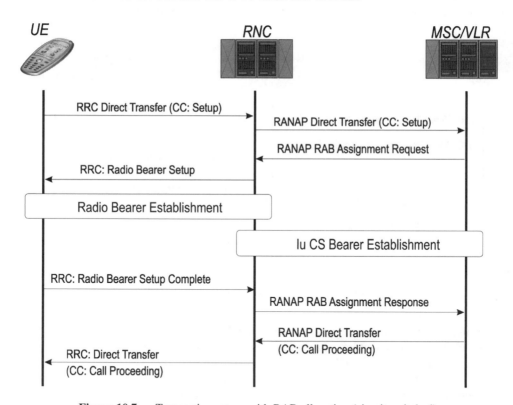

Figure 10.7.   Transaction set-up with RAB allocation (circuit switched)

In the case of circuit switched transaction the actual set-up information is delivered through RRC/RANAP direct transfer messages. These messages may carry in their payload the *Call Control (CC) set-up* message as shown in Figure 10.7. The CC set-up message identifies the transaction and indicates the QoS requirements, i.e. what kind of bearer is required for the service with the following parameters:

- Transaction identifier (TI)
- Stream identifier
- Traffic class
- Asymmetry indicator
- Maximum bitrate
- Guaranteed bitrate

With the TI value the UE and the CN node are able to separate calls from each other. Each call has its own TI value. The stream identifier recognises the bearer used for this call. If the stream identifier value does not exist yet, it means that the CC protocol in UE likes to establish a new bearer. The stream identifier value could be in use already. In that case the CC protocol likes to establish a multicall using an existing bearer given by the stream identifier.

Upon receiving the CC set-up message the MSC/VLR starts some activities. At first, the MSC/VLR checks whether the UE and current subscription have rights to perform the requested operation. If yes, the MSC/VLR starts RAB allocation by assigning a unique RAB ID and requesting a RAB with given QoS parameters to be set-up by sending the message *RANAP RAB assignment request* over the Iu interface.

When the RNC receives the RANAP RAB assignment request it starts activities for radio bearer allocation by first checking if there are enough resources available to satisfy the requested QoS. If yes, the radio bearer is allocated according to the request. If not, the RNC may select either to continue the allocation with lowered QoS values (for instance the maximum bitrate for the bearer could be lowered) or it may select to queue the request until radio resources become available. These special cases require some additional signalling not shown in Figure 10.7.

The RNC informs the UE about the bearer allocation by sending the message *RRC radio bearer set-up* to the UE. When the UE receives this message it is able to combine the information it originally sent to the network in its CC set-up message with the received radio bearer identifier. After this the UE can route the user data traffic (user plane) to the correct radio bearer. As soon as the UE is able to receive data from the new radio bearer it acknowledges this by sending the *RRC radio bearer set-up complete* message to the RNC. The RNC must also establish a Iu bearer for the new transaction.

After this the RNC indicates that RAB has now been allocated by sending the message *RANAP RAB assignment response* to MSC/VLR. If the RNC made exceptions from the QoS values requested by the MSC/VLR, the changes are indicated in this message.

Now the RAB is established and the procedure continues on the CC protocol level.

In case the RAB is to be established for a packet switched transaction, the procedure looks pretty much the same (see Figure 10.8) and the differences are within the messages. Instead of the CC protocol the UE now uses the SM protocol. The UE sends the *SM activate PDP context request* message to the network as an RRC direct transfer message and the RNC relays this as a RANAP direct transfer message to the SGSN in the CN.

Radio bearer allocation is similar to circuit switched traffic except that the parameters describing the QoS properties of the bearer are different. For example, guaranteed bitrate is an essential parameter in the case of a packet switched RAB but is not significant with circuit switched RABs.

After the RAB has been established the CN domain confirms the packet session establishment by sending an *SM activate PDP context accept* message. In terms of Session Manage-

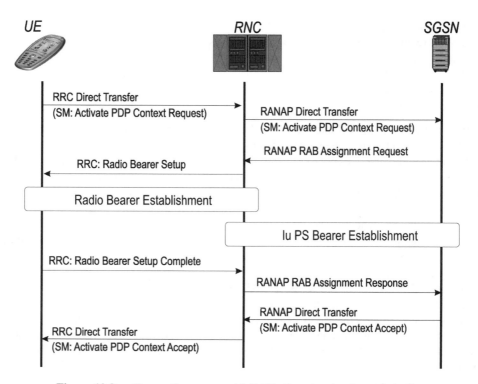

Figure 10.8. Transaction set-up with RAB allocation (packet switched)

ment (SM) this means that the SM state is now active and the UE and the CN domain are able to exchange packet switched data.

10.1.6. Transaction

In the most typical cases this phase of the transaction contains an active user plane connection. Examples of this kind of transactions are given later in this chapter (see e.g. Section 10.4). However, if the RRC connection was requested for a pure signalling purpose like an MM activity, no user plane bearers are needed. Instead, a signalling connection to perform the MM activity is used here as shown for instance in Section 10.3.3.

10.1.7. Transaction Clearing and RAB Release

If the transaction had a user plane active the user plane is disconnected first. The disconnect procedure depends on the transaction type. Figure 10.9 shows how a circuit switched transaction is cleared. In this case the clearing is started by the UE but the CN can also start the user plane disconnection. More generally, either party can clear a normal circuit switched call and this does not depend on which party initiated the call.

When the user plane has been disconnected the system releases the RAB with a separate subprocedure. When this is done with *RANAP RAB assignment request* and *RANAP RAB assignment response* indicating release as Figure 10.9 shows, the signalling connection still

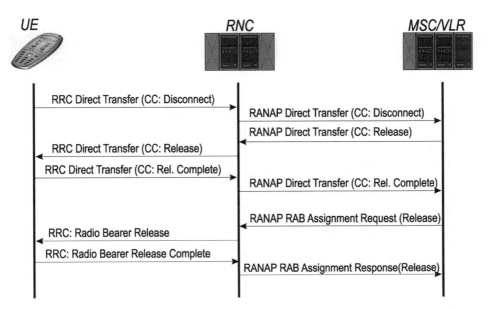

UE   RNC   MSC/VLR

RRC Direct Transfer (CC: Disconnect)

RANAP Direct Transfer (CC: Disconnect)

RANAP Direct Transfer (CC: Release)

RRC Direct Transfer (CC: Release)

RRC Direct Transfer (CC: Rel. Complete)

RANAP Direct Transfer (CC: Rel. Complete)

RANAP RAB Assignment Request (Release)

RRC: Radio Bearer Release

RRC: Radio Bearer Release Complete

RANAP RAB Assignment Response(Release)

Figure 10.9.   Transaction clearing with RAB Release (Circuit Switched)

remains. The UE still has an RRC connection to the UTRAN and any other RABs for the same UE may still exist. In the general case with a single *RANAP RAB assignment request* message the CN may cause a number of RABs to be created, deleted or modified.

If all the RABs between the UE and CN domain are released at the same time and also the related radio resources are deallocated, this can be done by using *RANAP Iu release command* as shown in Figure 10.10. Upon receiving this message the RNC starts *RRC radio bearer release* over the Iub interface. Also all possible radio bearers established through DRNCs are released.

When the radio bearers are released the RRC connection is also released. After all radio bearers and the RRC connection are released the RNC informs the CN domain with a *RANAP Iu release complete* message.

In the case of packet switched connections the user plane and RAB(s) are cleared in the same way as in the case of circuit switched connections. Again the differences are in the message parameter level. Figure 10.11 illustrates the case where the packet switched connection is closed by deactivtion of the PDP content.

The closing of a packet switched connection is one of the SM protocol tasks where the SM changes its state from active to inactive. The signalling procedure used for this state transition is PDP context deactivation. When the UE desires to do this, it sends an RRC direct transfer message containing *SM deactivate PDP context request* to the SRNC. The SRNC further relays it as a RANAP direct transfer message to the CN PS domain.

PDP context deactivation request indicates to the SGSN that the RAB allocated for this packet switched connection is not needed any more and thus it should be released. RAB release is done in the same manner as in the case of circuit switched connection.

When the RAB has been released the SGSN indicates that packet switched connection is now deactivated by sending the *SM deactivate PDP context accept* message to the UE.

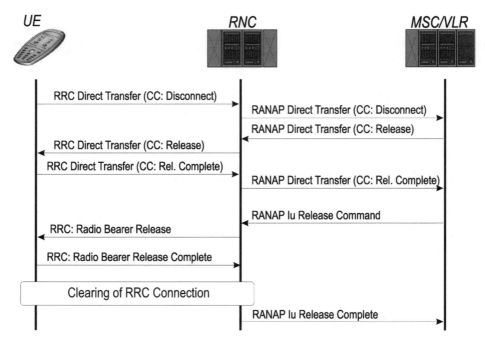

Figure 10.10.   Transaction clearing with Iu release (circuit switched)

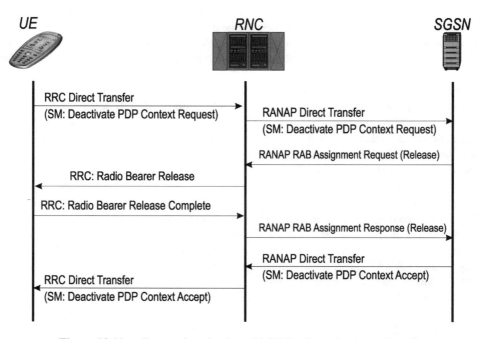

Figure 10.11.   Transaction clearing with RAB release (packet switched)

After this procedure the UE still has an RRC connection with the network and it may have other circuit switched and packet switched connections open. If all of the packet switched connections are to be terminated and also related radio resources deallocated, the system uses the *RANAP Iu release command* like in the case of circuit switched connections (see Figure 10.12).

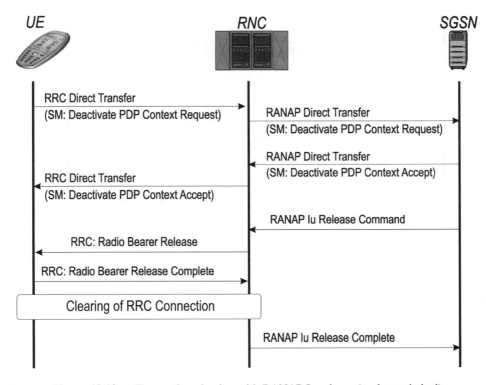

Figure 10.12.   Transaction clearing with RANAP Iu release (packet switched)

10.1.8. RRC Connection Release

The RRC connection release procedure, which is shown in Figure 10.13, is always started by the RNC. The RNC identifies which RRC connection is to be released and then sends this information to the UE in the message *RRC connection release*. The UE acknowledges that the RRC connection has been released by sending the message *RRC connection release complete*.

After this the RNC starts to clear the Iub interface resources. This is done by exchange of *NBAP radio link deletion* and *NBAP radio link deletion response* messages. When the radio link deletion is agreed on the NBAP level, the transport data bearer at the Iub interface is released. Any radio links established over a DRNC also has to be released by using the Iur subprocedures (not shown in Figure 10.13).

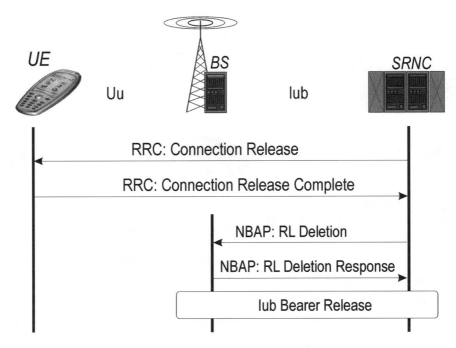

Figure 10.13.   RRC connection release

10.2. RRM Procedure Examples

As was described in Chapter 4, the RRM is a collection of algorithms needed to establish and maintain a good quality radio path between the RNC and the UE. Thus, basically every algorithm the RRM contains is present in any transaction the UTRAN is involved in. The RRM algorithms are handover algorithm, power control algorithm, admission control and packet scheduling as well as code management. From these, power control, admission control and code management are used "continuously", i.e. from the very beginning of the transaction and they are involved in every phase of a transaction. Packet scheduling is used in the context of packet switched transactions and handovers are applied for both circuit and packet switched transactions whenever needed.

The RRM entities located in UE and RNC communicate on radio resource management between themselves by using the RRC protocol. Therefore the establishment of an RRC connection, as described in Section 10.1.2, is a must for every transaction. However, it is important to keep in mind, that for any UE one RRC connection with UTRAN is enough no matter how many radio bearers are simultaneously open for various system-wide transactions.

In this subsection we present some examples of handovers. The handover type to be explained first is soft handover, which follows the principles presented in Chapter 4. After soft handover examples, we give an overview of the SRNS relocation procedure, where UE is not involved in the procedure. The last RRM procedure example is the inter-system handover case where the UE performs handover from WCDMA to GSM radio access.

10.2.1. Soft Handover – Link Addition and Link Deletion

When the UE has a service in use the RRC connection with the UTRAN exists and is active. In this case, the UE continuously measures the radio connection and sends measurement reports to the SRNC. The content of the measurement report was presented in Chapter 4 in Section 4.1.3.1.

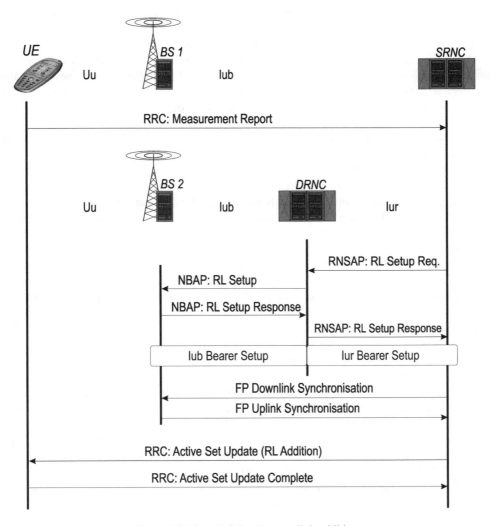

Figure 10.14.   Soft handover – link addition

The handover algorithm located in the SRNC averages and investigates the contents of the received measurement reports. Based on the results the SRNC realises that the UE has measured a cell located in BS 2 to have radio conditions fulfilling handover criteria defined in the SRNC. Based on the radio network information stored in the SRNC database the SRNC finds out that the target cell in BS 2 does not belong to the same RNS.

The SRNC starts arrangements on the UTRAN side by requesting through the Iur interface the DRNC to set-up a new radio link. This is done by sending a *RNSAP RL set-up request* message. This triggers the DRNC to establish a radio link over the Iub interface between the DRNC and BS 2 with NBAP protocol exchange.

After these steps the Iub and Iur bearers are established and frame protocols are synchronised in the downlink and uplink directions between the SRNC and BS 2. The frame protocols in the Iub and Iur interfaces implement the radio network user plane and carry actual user data flow. In this example we assumed that the used service is a voice call. Thus, the frame protocol used in this example is the Iub/Iur Dedicated Channel (DCH) frame protocol.

When the SRNC has received the FP uplink synchronisation, it sends a *RRC active set update* message to the UE. In this message the SRNC indicates to the UE that a new radio link has been added to the active set of the connection through the cell located in the BS 2 and that the connection can be taken into use. The UE acknowledges this by responding with *RRC active set update complete*.

From the bearer architecture point of view, this example illustrates a situation where the UE has active, circuit switched RAB. This RAB is realised within UTRAN by the Iu bearer located between the SRNC and CN and one radio bearer going from the SRNC to the UE through a cell located in BS 1. The procedure described in Figure 10.14 is, in the sense of QoS and bearers, a case where the SRNC adds an additional radio bearer to the existing connection. The RAB and Iu bearer remain unchanged. When the frame protocols are synchronised the SRNC performs RAB mapping to the radio bearers. This situation and related protocols are illustrated on a general level in Chapter 4 in Figure 4.29.

When the UE moves in the network during the transaction, it comes to the point where the SRNC finds out from the received measurement reports that a radio connection carrying the radio bearer through a cell located in BS 2 does not fulfil the criteria set to the radio connection any more. When this happens, the SRNC indicates to the UE that this particular radio connection can be removed from the active set. This is done by sending again an *RRC active set update* message to the UE indicating the radio connection to be removed. The UE acknowledges the radio connection removal by sending an *RRC active set update complete* message to the SRNC.

Upon receiving the UE's acknowledgement the SRNC can now start the radio link deletion between itself and the BS 2. This is done by using a *RNSAP RL deletion request* over the Iur interface, which in turn triggers the DRNC to delete the radio link in the Iub interface with NBAP protocol exchange. When the radio link has been deleted both in Iub and Iur, the related Iub and Iur bearers are also released. The message flow related to a soft handover with a radio link deletion is illustrated in Figure 10.15.

From the bearer architecture point of view this radio link deletion example illustrates the case, where the SRNC removes one radio bearer but RAB between the CN and UE remains.

10.2.2. SRNS Relocation – Circuit Switched

The role of Serving RNC (SRNC) means the RNC maintains the Iu bearer and performs RAB mapping to the radio bearers with respect to a single active UE. Since the radio link components of the radio bearers may continuously change due to soft handovers (refer to Section 10.2.1), the SRNC may end up in the situation, where it does not manage any radio links with its own BSs directly over the Iub any more. When this happens, the RAB mapping function-

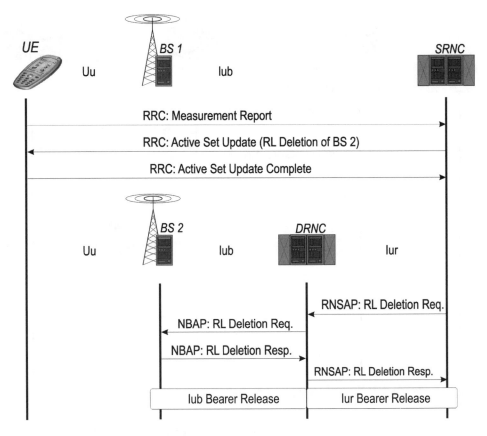

Figure 10.15.   Soft handover – link deletion

ality should be changed to another RNC which is in a better position to perform it. Since the RAB mapping requires the existence of the Iu bearer in the same RNC, the CN connection will change, too.

The RRM procedure where the Iu bearer termination is changed from one RNC to another (and at the same time the radio bearers) is called SRNS relocation. The message flow of this procedure is illustrated in Figure 10.16. In this figure, the RNC1 is the original SRNC and the RNC2 is the RNC, which will adopt the SRNC functionality.

The SRNS relocation procedure can be done in two different ways: either the UE is actively involved in the procedure or it is not. The example presented here illustrates the situation where the UE is not involved.

When the original SRNC (RNC 1) realises that the SRNC functionality should be transferred to a better RNC candidate (RNC 2), it starts this procedure by contacting the relevant CN domain with a *RANAP relocation required* message. This message contains the reason for the relocation, target RNS identification and UE classmark information. Based on the target RNS identification the CN domain is able to route this query further onto the target RNS. This re-routed query is the *RANAP relocation request* message. In addition to other relocation related information, the RANAP relocation request contains information about bearer

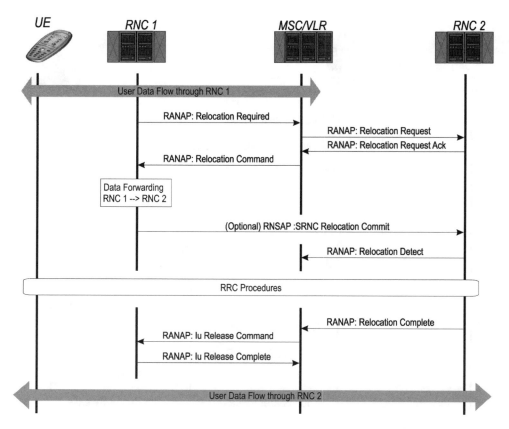

Figure 10.16.   SRNS relocation – circuit switched (UE not involved)

contexts with which the CN at the same time effectively moves the RAB and RRC connection end-point from the original SRNC (RNC1) towards the target SRNC (RNC2).

If the target RNC is able to provide resources to handle the incoming SRNC functionality it acknowledges the request back to the CN, which relays this acknowledgement to the source RNS. From the RNC1 point of view this acknowledgement is a command to start relocation, since everything should be now ready in the target RNC.

Upon receiving the RANAP relocation command the original SRNC (RNC1) starts to forward data to the target SRNC (RNC2). The expected SRNC (RNC2) realises the incoming data and informs the CN domain about the detection of SRNS relocation with the message *RANAP relocation detect*. Optionally, the source SRNC (RNC1) may send a *RNSAP SRNC relocation commit* message to the target SRNC (RNC2) over the Iur interface.

In the case of SRNC relocation where the UE is not involved, the UE is not aware of the message flow described above. If there is a need to perform any RRC procedures, they are performed after the RANAP relocation detect message. UTRAN may, for instance, send an *RRC UTRAN mobility information* message to the UE. With this message the UTRAN updates C-RNTI and U-RNTI and possibly other relevant MM information at the UE.

When all relevant RRC procedures are performed the new SRNC (RNC2) informs the CN that SRNC relocation procedure is now complete by sending the message *RANAP relocation complete*. This then triggers the CN to release resources related to the old SRNC (RNC1) by issuing a RANAP *Iu release* message, which is acknowledged by the old SRNC (RNC1).

Now the user data flows will go through the new serving RNC and, according to its role, it acts as the macrodiversity combination point for the UE and also controls all RRC activities for this UE.

10.2.3. Inter-System Handover from UMTS to GSM – Circuit Switched

The 3GPP specifications define that a handover between two radio accesses is a mandatory requirement for the system. From the UMTS point of view this means that the system must be able to perform handover between UTRAN and a GSM BSS.

Inter-system handover between UTRAN and GSM BSS is a special case of the SRNS relocation procedure as can be seen from Figure 10.17. Actually RANAP carries the same kind of information as the BSSMAP protocol in the GSM network and also much more. If the handover is performed from UTRAN to a GSM BSS, the UTRAN side messages are exactly the same as in the case of SRNS relocation but the contents of those messages vary. For example, the message *RANAP relocation required* contains the target RNC ID in the SRNC relocation case. When the target is on the GSM side, the target RNC ID is replaced with cell global identity, which is a parameter more familiar to the GSM BSC.

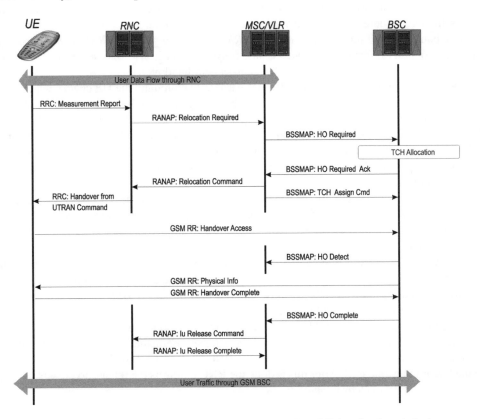

Figure 10.17.   Inter-system handover from UMTS to GSM – circuit switched

Unlike the SRNS relocation, the inter-system handover is a procedure where the UE is always involved, since the UE must perform Radio Resource (RR) activities in order to access the GSM BSS; note that the GSM BSS is not able to handle any kind of context information valid in UTRAN.

The UE must have the possibility to measure the GSM cells surrounding the current UTRAN cell(s). This is made possible in UTRAN by using the slotted mode procedure (see Section 4.1.3.1). Slotted mode gives the UE some time to perform measurements on the GSM band and thus detect any possible handover candidate(s). Information about these GSM cell candidates is delivered to the RNC with measurement reports just like that of the UTRAN cells.

As soon as the RNC realises that a GSM cell is the best candidate and no suitable UTRAN cells are available, the SRNC starts to require relocation from the CN. The CN checks the contents of the message *RANAP relocation required* and finds out that the target cell for this handover belongs to a GSM BSS. Thus, the CN sends the message *GSM BSSMAP handover required* to the target BSC on the GSM network side.

This triggers the target BSC to set up Traffic Channel (TCH) for the connection to be handed over to it. When the TCH allocation has been successfully performed within the GSM BSS the handover request is acknowledged back to the CN. The CN relays this message back to the SRNC as a *RANAP relocation command* message.

The RANAP relocation command starts the relocation procedure within the SRNC. Since this is the case where the UE is involved the SRNC commands the UE to perform inter-system handover by sending the message *RRC handover from UTRAN command*. This message contains information about the target system and it also may carry in its payload any additional information related to the inter-system handover.

Upon receiving the RRC handover from the UTRAN command the UE checks whether the message indicated any time to perform the handover (default is immediately) and starts handover actions accordingly. Since the target radio access is GSM, the UE sends a *GSM RR handover access* message to the target cell in the GSM BSS. When the GSM BSS target cell detects this, it indicates to the BSC that the UE is accessing the GSM BSS. The GSM BSC in turn informs the CN about the incoming UE by sending the message *GSM BSSMAP handover detect*.

As an acknowledgement to the GSM RR handover access the UE receives *GSM RR physical info* from the GSM BSS cell. This message contains information with which the UE is able to start to use the GSM radio access, for example, channel descriptions are sent here to the UE.

Finally, when the UE has successfully accessed the GSM BSS cell, the UE sends a *GSM RR handover complete* to the GSM BSC. The BSC in turn relays the same information to the CN thus indicating that the UE has now entered into the GSM BSS and the inter-system handover is successfully completed in this respect.

Since the UE does not use UTRAN resources any more, all resources related to the UE can be released and the CN issues a *RANAP Iu release* command. The RANAP Iu release command in turn triggers the RNC to clear the RRC connection and all the resources related to the UE are thus released. After this is done the RNC confirms the release to the CN with the message *RANAP Iu release complete*.

10.3. MM Procedure Examples

When compared to GSM, the MM behaves mostly the same way in UMTS but there are, however, some differences. MM as an entity covers all procedures, methods and identities required to maintain knowledge about the UE's location when it is moving in the network. Due to the existence of packet switched transactions a state model for MM was introduced, please refer to Chapter 5.

In GSM networks MM is completely handled between the MS and NSS. In UMTS networks most of the MM functions are equally handled between the UE and the CN but not all. The RNC partially handles the UE's movement within the radio access network using RRC procedures for this purpose. The MM activities handled by the RNC are cell and URA (UTRAN Registration Area) updates

10.3.1. Cell Update

During a circuit switched transaction, the CN knows the location of a UE with the accuracy of a location area and the RNC knows the location of the UE with the accuracy of a cell or – to be more precise – cells if the UE is in soft handover state. The cell information is continuously changing during the transaction. The pace of change depends on radio network conditions and whether the UE is moving and with what speed.

In the case of packet switched transactions the location information is distributed in such a way that the CN PS domain knows the location of the UE with the accuracy of location area and routing area and the RNC knows the location of the UE with the accuracy of a cell. Because packet switched connection is discontinuous in nature, the cell information in the RNC will actually be out-of-date when the data connection between the UE and the CN is temporarily closed. Hence, the packet switched connection is enabled from the UE and the CN point of view but the RNC is not able to route the data packets in this state.

If the UE desires to send data packets and the current cell is different than it was earlier, the UE performs an *RRC cell update* procedure (Figure 10.18) to inform the RNC about the current cell. After this basically any RRC activity is possible, for instance, a radio bearer can be established.

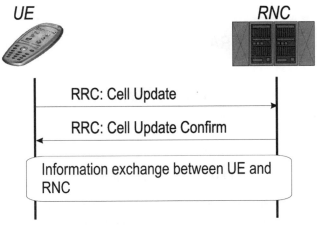

Figure 10.18.   Cell update procedure

If, on the other hand, it is the CN PS domain that has some data packets to be sent to the UE, the CN PS domain first issues a RANAP paging to make the RNC send a RRC paging Type 1 to the UE. Upon receiving this the UE checks whether the cell update procedure is needed. If yes, the UE performs cell update and after that it is ready to start packet reception.

To summarise, the various reasons for a UE to perform a cell update are as follows:

- Cell reselection
- Periodic cell update
- Uplink data transmission
- Paging response
- Re-entered service area
- Radio link failure
- Unrecoverable RLC error

In association with a cell update procedure RNC may allocate a new RNTI to the RRC connection.

10.3.2. URA Update

The RNC maintains registration of the current URA for each UE. As described in Chapter 5 a URA consists of number of cells belonging to either one RNC or several RNCs. Like LA and RA, the URA is one of those identities the UE stores in USIM. When switched on, the UE continuously monitors the received URA identity. If the received URA differs from the one stored in USIM the UE performs an *RRC URA update* procedure (Figure 10.19).

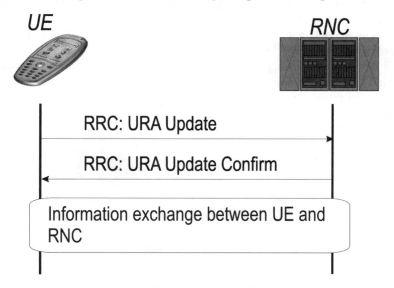

Figure 10.19. URA update procedure

Other conditions, which trigger a URA update procedure, are, when the periodic URA update timer has expired or the UE has re-entered a network service area.

As in the case of cell update, the RNC changes the RNTI allocated for the UE.

10.3.3. Location Update to the CN CS Domain

Figure 10.20 illustrates the signalling message flow for a location update performed between the UE and the CN circuit switched domain.

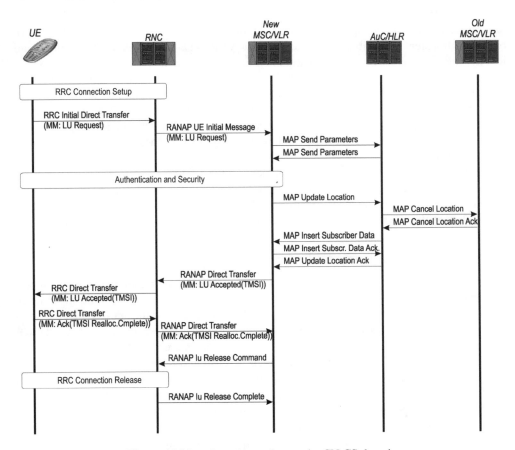

Figure 10.20.   Location update to the CN CS domain

If the UE does not have RRC connection with UTRAN the RRC connection is established first. Actual Location Update (LU) is triggered by LA identity. The UE has stored the LA in the USIM and if the LA identity received from the system information of the camped cell is different than the one stored in USIM the UE starts the location update procedure in order to inform the CN about its current location.

The *MM LU request* message is carried in the RRC initial direct transfer envelope to the RNC where it is relayed to the CN within a RANAP UE initial message. The LU request message contains an old LA identity and a new LA identity and also the subscriber identification is added. In the normal case, this identity is a TMSI number allocated previously by the MSC/VLR.

The VLR checks whether it has authentication vectors for this subscriber. If they are not present, the security parameters are fetched from the AuC, which sends some predefined amount of them in the *MAP send parameters* message to the VLR. Upon receiving these parameters the VLR is able to start the access level security procedures for authentication and ciphering.

If the old LA identity and the new LA identity differ from each other the VLR informs the subscriber's HLR about the new location. The HLR cancels the old location of the UE with a *MAP cancel location* message and starts a *MAP insert subscriber data* operation. This MAP operation transfers subscriber profile to the new VLR. The subscriber profile contains all identity and service provision information and possible restrictions concerning their use.

When the VLR has updated itself with the subscriber and his/her location information the CN domain informs the UE of the successful completion of the procedure by sending am *MM LU accepted* message. This message contains also new a TMSI for the subscriber. The RNC relays this information with RRC direct transfer to the UE. The UE acknowledges the LU acceptance and a new TMSI value by sending the message *MM TMSI reallocation complete* to MSC/VLR.

When the VLR has received acknowledgement for the TMSI reallocation it starts to release the signalling connection. The Iu interface is released with the *RANAP Iu release command* and upon receiving this the RNC releases the RRC connection. When completed the RNC acknowledges the release by sending the *RANAP Iu release complete* message to MSC/VLR.

10.3.4. Routing Area Update to the CN PS Domain

As already discussed in Chapter 5, the CN packet switched domain maintains its own mobility registrations for UEs. The PS domain surely recognises LA but the more meaningful registration in this context is Routing Area (RA). The UE has stored the latest RA where it has performed RA update. From the system information broadcasted by the camped cell the UE receives its current RA identity. If this differs from the one stored in USIM the UE starts the Routing Area Update (RAU) transaction, which is illustrated in Figure 10.21.

The UE first sends within a direct transfer message the *GMM RAU request* message. This message contains the old RA and new RA identifier values. This message arrives through the Iu interface to the new SGSN. The new SGSN knows the "MM environment", i.e. the neighbouring RAs and their relationship to other SGSNs in the network. Based on this information the new SGSN is able to determine the old SGSN and it requests information about the subscriber from there with the *GTP-C SGSN context request* message. In other words, the new SGSN finds out whether the UE performing RAU has already an existing PDP context.

After this the new SGSN may request the subscriber's AuC to provide authentication vectors. The AuC calculates those and returns within a *MAP send parameters* message a set of vectors to the new SGSN. Now the new SGSN is able to start authentication and security control for the UE over UTRAN.

When authentication and security activities have been successfully completed the new SGSN informs GGSN by sending a *GTP-C update PDP context* message to the GGSN. The new SGSN informs the GGSN that the SGSN has now been changed and the PDP context information has been changed accordingly.

Now the new SGSN updates location information with the HLR with a *MAP update location* message. Upon receiving this, the HLR cancels the old location of the UE from the old SGSN. When the old location has been cancelled the HLR starts to transfer the subscriber profile to the new SGSN with a *MAP insert subscriber data* operation.

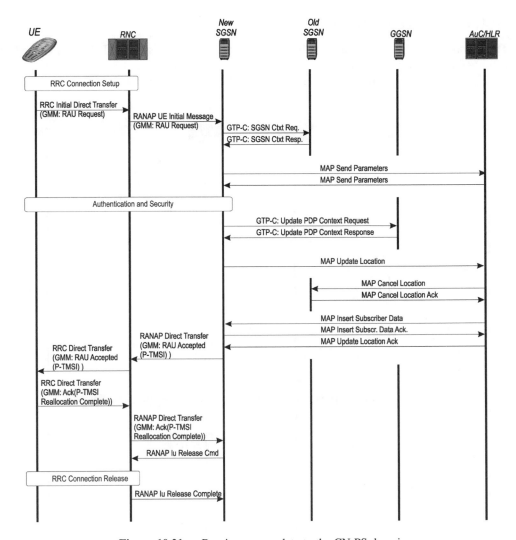

Figure 10.21.   Routing area update to the CN PS domain

When the subscriber profile has been updated in the new SGSN it sends a *GMM RAU accepted* message to the UE. This parameter carries as a parameter the new P-TMSI number to the UE. The UE stores this P-TMSI on its USIM and acknowledges the receipt of it by sending a GMM acknowledgement to the SGSN.

Now the RA update is completed the new SGSN releases the signalling connection used on this transaction.

The message flow in Figure 10.21 presents RAU in case the SGSN changes. SGSN is not changed in every RA update. If the SGSN remains the same the procedure is simpler since the MAP dialogues between the SGSN and HLR are not needed.

10.4. CC Procedure Example

The Figure 10.22 illustrates the signalling message flow of a UE terminated circuit switched call coming from a PSTN (Public Switched Telephone Network). The PSTN signalling and signalling used between the MSC nodes is assumed to be ISUP (ISDN User Part).

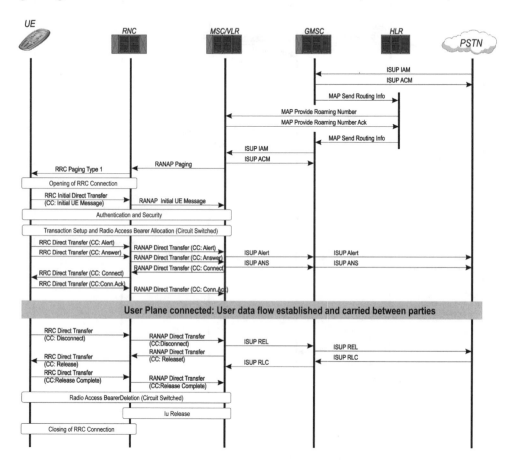

Figure 10.22. Circuit switched call – UE terminating – UE clears

The call first enters the gateway MSC from the PSTN by a *ISUP IAM* (Initial Address Message). This message contains the MSISDN number of the called UMTS subscriber. At the same time this MSISDN number identifies the service desired to be used. To set up the circuit switched path between the GMSC and PSTN the GMSC responds with an *ISUP ACM* (Address Complete Message).

The GMSC requests the HLR to provide routing information for the subscriber by sending a *MAP send routing information* message. In this message the GSMC sends to the HLR the received MSISDN number, from which the HLR is able to find out the IMSI of the subscriber. It also finds the latest reported location of this subscriber with the accuracy of the current MSC/VLR address. The HLR then issues a *MAP provide roaming number* request message to that MSC/VLR, which allocates and returns a roaming number (MSRN) for circuit switched path connection. When receiving this the HLR relays the MSRN to the GMSC. The MSRN contains the necessary information for call routing purposes, since every MSC/VLR in the network allocates MSRNs from a certain numbering space and this numbering space is recognised in the CC entity of the GMSC. The GMSC uses the received MSRN for call routing to the MSC/VLR with *ISUP IAM* and *ISUP ACM* messages as described earlier.

The MSC/VLR recognises the MSRN it allocated and determines that this call is to be terminated to the UTRAN under itself. Thus, it generates *RANAP paging* messages to the RNCs maintaining the cells, which belong to the LA, where the addressed UE has performed the latest location update. The RNCs send *RRC paging Type 1* messages to the Uu interface and the paged UE responds by starting the RRC connection establishment procedure.

Actual response to paging is carried by a *RANAP UE initial message* and upon receiving this the CN starts authentication and security activities for the call. If the security activities are successful the CN starts transaction set-up by informing the UE with s *CC set-up* message about an incoming call and its nature.

If the UE accepts this call the CN starts RAB allocation. When the RAB is allocated the UE alerts the subscriber and informs the network that the subscriber has been alerted by sending a *CC alert* message. This information is carried through the network(s) up to the calling subscriber exchange. When the subscriber answers the CC connect procedure is triggered over the UTRAN and answering information is carried over the network(s) with *ISUP ANS* (Answering Message).

Now the circuit switched call path is opened, charging starts, related voice codecs start and the system is ready to transfer user data flow since the user plane connection is now open.

It is further on assumed in this example that the UE clears the connection. A three-way clearing procedure is used by exchanging the CC *disconnect, release* and *release complete* messages over the UTRAN. On the ISUP side the same thing is taken care of by the messages *ISUP REL* (release) and *ISUP RLC* (release complete). These messages stop charging and clear the connection between nodes. After the circuit switched path is cleared the CN releases the RAB and RRC connection by invoking the Iu release procedure.

10.5. Packet Data Example

Figure 10.23 shows a small sample of what happens during a packet session. Naturally we also assume that the UE is moving in the UTRAN area continuously.

Under these circumstances the user application in the UE likes to send data packets to the network (uplink). Since the UE is continuously moving it first has to perform a cell update. By doing this the UE ensures that its location information with cell accuracy in the SRNC is valid.

After the cell update the UE is able to request PDP context activation through the current cell and also the RAB can be allocated for the packet data transfer. To see a complete message flow for this procedure, please refer to Figure 10.8.

When the packet session is open the UE transmits uplink data packets and they at first arrive at the SRNC over the radio links forming a radio bearer for this session. Packet traffic is then tunnelled between the RNC and the GGSN with the GTP-U protocol and physically this tunnel consists of two parts, the Iu bearer over the Iu interface and the PS domain backbone bearer between the SGSN and GGSN. The tunnel ends in the GGSN, from where the user data

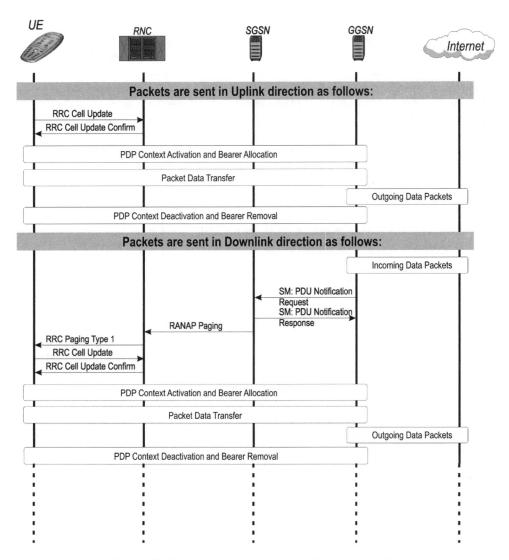

Figure 10.23.   Packet data transfer – uplink and downlink

packets are routed further on to an external packed data network according to a method described in the PDP context parameter connection type. This parameter may have values like PPP (Point-to-Point Protocol) or X.25. When there is nothing to transfer any more the PDP context is deactivated and RAB is released. For a complete message flow of this deactivation, please refer to Figure 10.11.

In the downlink direction the GGSN may later on realise that there are packets coming in from other networks. This triggers the SM PDU (Packet Data Unit) notification procedure which leads to network-initiated PDP context activation. Upon receiving the *SM PDU notification request* message the SGSN generates paging to the RNCs controlling the RA where the UE has performed the latest RA update. The RNCs deliver paging further on towards the

UE as RRC paging Type 1. Since the UE is continuously moving it has to perform cell update again in order to record valid cell information in the RNC.

   After the cell update the PDP context is activated, RAB is allocated and the data packets start to flow from GGSN to UE. If the UE has now any data packets to be sent they can be sent at the same time. The data packets transferred in the uplink direction are treated as described earlier. When there isn't any data to be transferred the PDP context is deactivated again and the RAB is released.

Bibliography

This bibliography includes some literature, which an interested reader may find useful when more information on the following topics is needed

- Radio communications, especially CDMA systems
- 3GPP technical specifications
- ATM networks and protocols
- IP networks and protocols
- Mobile positioning
- WAP protocols and services
- Security algorithms and protocols

The material listed below can be used as a starting point for more reading.

1. Technical Specifications

In the modern world the publicly available technical specifications can be easily found on the World Wide Web. Please give the following URLs to your web browser:

- 3GPP technical specifications: http://www.3gpp.org

The 3GPP specifications referred to in this book are mostly from the release 99 set of specifications. The references made in the book chapters and given in the numbering format TS 12.345 can be used for browsing the list of specifications on the 3GPP website.

- Internet Engineering Task Force (IETF) specifications: http:// www.ietf.org

The IETF specifications referred to in this book are indicated with their RFC (Request For Comments) numbers and can be found on the IETF website with those numbers.

- Wireless Application Protocol (WAP) Forum specifications: http://www.wapforum.org

The WAP Forum website includes WAP technical pages from which all the current WAP protocol specifications can be retrieved.

2. Other Literature

Black U. ATM, volume II: Signaling in broadband networks. Englewood Cliffs, NJ: Prentice Hall, 1998, 224 pp.

Black U. ATM, volume I: Foundation for broadband networks. Englewood Cliffs, NJ: Prentice Hall, 1999, 450 pp.

Comer DE. Internetworking with TCP/IP, volume I: Principles, protocols and architecture, 4th edition. Englewood Cliffs, NJ: Prentice Hall, 2000, 755 pp.

Doraswamy N, Harkins D. IPSec: the new security standard for the Internet, intranets, and virtual private networks. Englewood Cliffs, NJ: Prentice Hall, 1999, 216 pp.

Ericsson Telecom AB. Understanding telecommunications I-2. Studentlitteratur AB, 1997, 493 + 677 pp.

Glisic S, Vucetic B. Spread spectrum CDMA systems for wireless communications. Boston, MA: A-H Publishers, 1997, 383 pp.

Händel R, Huber MN, Schröder S. ATM networks, concepts, protocols, applications, second edition. Reading, MA: Addison-Wesley, 1994, 287 pp.

Heine G. GSM networks: protocols, terminology and implementation. Boston, MA: Artech House, 1998, 416 pp.

Holma H, Toskala A. WCDMA for UMTS, revised edition. Chichester: Wiley, 2001, 313 pp.

Lee WCY. Mobile cellular telecommunications; analog and digital systems, second edition. New York: McGraw-Hill, 1995, 664 pp.

Manterfield R. Telecommunications signalling. UK: The Institution of Electrical Engineers, 1999, 435 pp.

Menezes AJ, van Oorschot PC, Vanstone S. Handbook of applied cryptography. Boca Raton, FL: CRC Press, 1996, 780 pp.

Mouly M, Pautet M-B. The GSM system for mobile communications. France: Mouly M, Pautet, MB, 1992, 701 pp.

Ojanperä T, Prasad R. Wideband CDMA for third generation mobile communications. Boston, MA: Artech House, 1998, 439 pp.

Rantalainen T, Spirito MA, Ruutu V. Evolution of location services in GSM and UMTS networks. Proceedings of the Third International Symposium on Wireless Personal Multimedia Communications (WPMC 2000), November 2000, Bangkok, pp. 1027–1032.

Redl SH, Weber MK, Malcolm WH. An introduction to GSM. Boston MA: Artech House, 1995, 379 pp.

Spirito MA. Mobile stations location estimation in current and future TDMA mobile communication systems. Ph.D. Thesis. Politecnico di Torino, Facolta di Ingegneria, 2000.

Stallings W. Data & computer communications. Englewood Cliffs, NJ: Prentice Hall, 2000, 810 pp.

Viterbi AJ. CDMA principles of spread spectrum communications. Reading, MA: Addison-Wesley, 1995, 254 pp.

Index